D1165479

THE
OXFORD ENGINEERING SCIENCE
SERIES

General Editors

L. C. WOODS, W. H. WITTRICK, A. L. CULLEN

PLASMA
PHENOMENA
IN GAS
DISCHARGES

BY

RAOUL N. FRANKLIN

CLARENDON PRESS · OXFORD

1976

Oxford University Press, Walton Street, Oxford OX2 6DP

OXFORD LONDON GLASGOW NEW YORK
TORONTO MELBOURNE WELLINGTON CAPE TOWN
IBADAN NAIROBI DAR ES SALAAM LUSAKA ADDIS ABABA
KUALA LUMPUR SINGAPORE JAKARTA HONG KONG TOKYO
DELHI BOMBAY CALCUTTA MADRAS KARACHI

ISBN 0 19 856113 X

© OXFORD UNIVERSITY PRESS 1976

FILMSET IN NORTHERN IRELAND
AT THE UNIVERSITIES PRESS, BELFAST
PRINTED IN GREAT BRITAIN
AT THE UNIVERSITY PRESS, OXFORD
BY VIVIAN RIDLER
PRINTER TO THE UNIVERSITY

PREFACE

THE SUBJECT of plasma physics has enjoyed two periods of rapid growth. The first, the naissance, around 1930 occurred when Langmuir appropriated the word to describe the fluid properties of an assembly of charged particles. The second, the renaissance, in the 1950s, was due to the possibility of obtaining cheap plutonium and later of almost limitless power from nuclear fusion.

During this second period the high temperatures required led to the visualization of high degrees of ionization, that is situations in which atoms were stripped by collisions of all their electrons leaving the nuclei bare. At this time it was frequently inquired of any experimental situation whether (a) the dimensions were large in comparison with the Debye length or (b) the collision processes were dominantly between charge particles or alternatively the degree of ionization was high? The implication of these questions was that unless the answers were 'yes' the measurements were of limited value as a contribution to the subject of plasma physics.

Plasma phenomena in gas discharges brings together developments that have gone on over the past decade or so which indicate that, while for the positive column of a low pressure gas discharge the answer to (b) above is in the negative, significant contributions to the understanding of plasma processes have been made under such conditions. While much of this work has been aimed at making measurements of the many waves which can propagate in a plasma it has been necessary first to obtain detailed knowledge of the steady state.

Although considerable advances have been made in the understanding and methods of treating processes in the positive column, if any reader feels that the field is completely worked out, may I refer him to the pioneer work of L. F. Richardson† in connection with human psychological behaviour.

Contributions to research in plasma physics have been made from widely scattered research groups in a number of countries notably at Stanford, Caltech, M.I.T., Brussels, and Prague, apart from those I have been closely associated with in Oxford and at the Culham Laboratory.

This book is intended to complement the textbook *Ionized gases* by Hans von Engel and the *Handbuch der Physik* article by Gordon Francis; it is based on lectures given in recent years to the postgraduate M.Sc. course in plasma physics at Oxford University, but the order of presentation has been governed by the particular aim set out above. It is also

† *Proc. R. Soc.* **162A,** 293 (1937) and **163A,** 380 (1937).

written in the knowledge that a book by R. W. Motley on the generation of contact-ionization plasmas (Q-machines) and measurements made in them was in preparation.

The list of references quoted, while extensive, is by no means exhaustive; in some cases an attempt at attribution is made; in others, particularly as regards experimental work, the tendency has been to quote the most complete or most generally accessible. Questions have not been given at the end of each chapter, but, in so far as steps have been omitted in derivations and some results quoted rather than derived, the student will inevitably set himself exercises in following the text in detail.

It is appropriate that I should pay tribute here to those who introduced me to the mysteries which I have endeavoured to make clear, namely Hans von Engel and the late Gordon Francis. I would also like to thank colleagues who have stimulated the development of my ideas, particularly Professor L. C. Woods and Drs J. E. Allen and J. R. Forrest and, for his invaluable comments on the manuscript in preparation, Dr. P. D. Edgley.

Finally, I would like to thank those authors whose results have been reproduced for their cooperation and also the publishers of journals who have given permission.

R. N. F.

Oxford
December, 1975

CONTENTS

INTRODUCTION

THIS book is intended to cover those aspects of behaviour of the positive column of a gas discharge in the steady state which are capable of explanation and to use the information gained to aid the understanding of the details of the wealth of waves which can occur in such columns.

The first chapter describes the motion of charged particles under slightly collisional conditions, taking into account collisions with neutrals and with other charged particles, and also deals with the simpler aspects of collective behaviour. In later chapters it is convenient to describe collisions in terms of a collision frequency, and the extent to which it is valid to treat the collision frequency per unit gas density as a constant for electrons and ions moving in equilibrium with an electric field is examined.

Chapter 2 sets up and solves the two-fluid equations for the steady-state column in the approximation that the positive and negative particle densities are equal, and, except where kinetic theory is needed to obtain specific results, the two-fluid treatment is used throughout. Chapter 3 extends the treatment to consider the effects of an axial magnetic field, while Chapter 4 relaxes the assumption of charge neutrality and thus deals with boundary effects and the formation of space-charge sheaths. Chapter 5 brings together several effects which cause departure from the ideal steady-state behaviour outlined in the previous three chapters, including current limitation at low pressures, contraction at high pressures, the effects of negative ions, and the afterglow immediately after the cessation of an active discharge.

The subsequent chapters deal with various types of wave which have been unambiguously observed to propagate in the positive-column plasma. The range covered is electron plasma waves, Bernstein waves, surface waves, ion waves, ionization waves, drift waves, and Alfvén waves. The list is not exhaustive, and it is sometimes true that more definitive measurements have been made on other types of plasma. But the attempt has been made to bring a wide range of work together and present it coherently. Where conditions arise such that a wave grows spatially or temporally the instability that this implies has also been discussed.

SI units are used throughout.

1

BASIC CONCEPTS OF DISCHARGES AND PLASMAS

1.1. The motion of a charged particle in a gas

As a charged particle moves through a gas it makes collisions with the gas atoms or molecules. These collisions may be ones in which the kinetic energy is conserved (elastic collisions) or may result in excitation of the target to an electronic, vibrational, or rotational state. Taking elastic collisions only for the moment, these can be studied in detail to determine the scattering through a certain angle in a single collision as a function of incident electron energy. A great deal of such information is available now, see, for example, Massey and co-workers (1969). On the other hand, in some circumstances one is only interested in the effect of a statistical average of such collisions supposing some equilibrium state. The simplest such situation to envisage is one in which there is on average uniform motion in some direction arising from the influence of an electric field in that direction as indicated in Fig. 1.1. Under these conditions one can introduce the assumption of a constant mobility μ of the charged particle defined by

$$v_D = \mu E, \tag{1.1}$$

where v_D is the (average) drift velocity and E is the uniform electric field strength. One can seek experimental confirmation for such a concept, and measurements show that for charged particles of either sign there are regions of validity at low fields where this is experimental justification. Figs 1.2 and 1.3 give v_D as a function of E/p for electrons and for ions in their own gas, p being the gas pressure. Of course one is led immediately to an 'Ohm's law' in such a situation if there is no charge multiplication, since then the current density nev_D is proportional to voltage.

The limits of validity involve both the onset of inelastic processes, when the particle random energy is sufficiently large, and also difficulties when the drift velocity becomes comparable with the particle random velocity or, alternatively, when there is a significant and rapid transfer of directed motion to random.

One can enquire whether there is any model of the particle motion which would lead to (1.1) with μ constant, given that the equation of

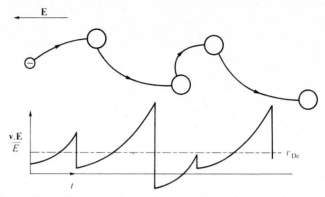

FIG. 1.1. Schematic motion of an electron in uniform electric field undergoing collisions with gas atoms and the corresponding time variation of the velocity component in the field direction. The mean value of this component is v_{De}.

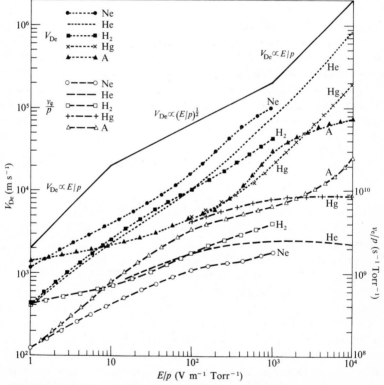

FIG. 1.2. Measured drift velocities v_{De} of electrons in some of the commoner gases as a function of reduced field E/p and the corresponding derived reduced frequency for momentum transfer v_{e}/p defined by $eE/mv_{\mathrm{De}}p$. Also shown for comparison are lines corresponding to a constant collision frequency $v_{\mathrm{De}} \propto E/p$ and to a field-dependent random energy given by the constant fractional energy-loss model described in §1.2.

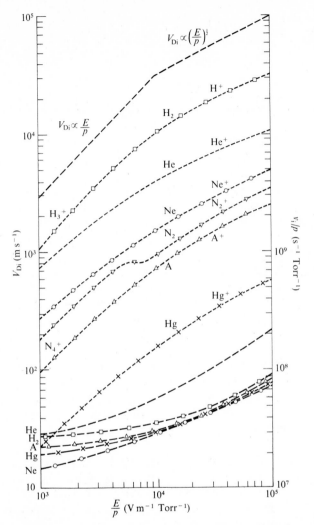

FIG. 1.3. Measured drift velocities v_{Di} of ions in the same gases as Fig. 1.2 as a function of E/p and the corresponding derived reduced for momentum transfer $\nu_i/p \equiv eE/Mv_{Di}p$.

motion giving the particle velocity $\mathbf{v}$ in the absence of collisions is

$$m\frac{d\mathbf{v}}{dt} = e\mathbf{E}.$$

Ignoring the initial acceleratory motion and taking a time average for the velocity over many collisions, the addition of a frictional term $m\mathbf{v}\nu$ to the left-hand side provides the required form, where ν is normally known as the collision frequency for momentum transfer. Figs 1.2 and 1.3 also show the effective collision frequencies for electrons and ions ν_e and ν_i defined by

$$\frac{\nu_e}{p} \equiv \frac{e}{m v_{De}} \cdot \frac{E}{p}$$

and

$$\frac{\nu_i}{p} \equiv \frac{e}{M v_{Di}} \cdot \frac{E}{p}$$

as functions of E/p, from which it can be seen that while they are slowly varying it is an approximation to take them as constant. Lines of slope 1 and $\frac{1}{2}$ are shown for comparison with the $v_D(E/p)$ curves since it will be shown in the next section that when inelastic processes are dominant one might expect $v_D \propto (E/p)^{\frac{1}{2}}$ rather than $v_D \propto E/p$. At very low fields, $E/p < 10^{-1}$ V m^{-1} Torr^{-1} for electrons, and two orders of magnitude larger for ions, collision frequencies are measured to be constant; but the range of values normal in positive columns, as is shown in Chapter 2, is $10 < E/p < 10^5$, so data is shown for such a range. There is a complication which occurs particularly at high pressures or lower values of E/p so far as measurement of ion drift is concerned, and that is a change in the identity of the ion which exists in collisional equilibrium with the parent gas so that for hydrogen the dominant ion becomes H_3^+ and for nitrogen N_4^+. In the latter case the effect is so marked as to produce a kink in the drift-velocity curve. Diatomic molecular ions form in the other gases at higher pressures, and a reasonable approximation is that their drift velocities are twice those of the atomic ion at the same value of E/p and consequently their collision frequencies reduced by a factor of approximately 4.

Since the collision frequency is defined in such a way as to give a constant drift velocity in the case of simple motion under a uniform electric field, any test of the usefulness of this model must go to more complicated field variations. Two such can be provided simply; one, not so sensitive, involves alternating electric fields alone, and the second requires the presence of a steady magnetic field. In the first case the equation of motion of an electron,

$$m\frac{d\mathbf{v}}{dt} + m\mathbf{v}\nu = \mathrm{Re}(-e\mathbf{E}\exp i\omega t), \tag{1.2}$$

where Re means real part and E is the peak field intensity of frequency ω. This predicts a power absorption per unit volume of the form

$$\frac{ne^2 E^2}{m} \cdot \frac{\nu}{\omega^2 + \nu^2},$$

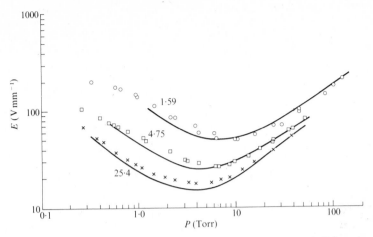

FIG. 1.4. Experimental points and theoretical curves for the field strength E for microwave breakdown in hydrogen as a function of gas pressure p corresponding to three different diffusion lengths L. The theory assumes a constant value for ν_e/p. (McDonald and Brown 1949).

where n is the electron density, and for variable ω the absorption has a broad maximum at $\nu \sim \omega$. Experimental measurements corresponding to this situation are shown in Fig. 1.4, where the electric field at *breakdown* at microwave frequencies, rather than the power absorption, is shown as a function of pressure. Under conditions such that the particle loss is independent of pressure and described by a frequency per particle ν_1, breakdown, i.e. charge multiplication, will occur when the power absorption is equal to the power loss $n\nu_1 e V_i$, where V_i is the ionization potential. Hence for a constant collision frequency the field would be expected to vary as

$$E^2 = \frac{m\nu_1 V_i}{e} \frac{p(\nu/p)^2 + \omega^2/p}{\nu/p}.$$

The theoretical curves are for $\nu/p = 5\cdot9 \times 10^9\,\mathrm{s}^{-1}\,\mathrm{Torr}^{-1}$, the value appropriate to hydrogen in the range of effective fields used.

1.2. Inclusion of the effects of inelastic collisions

Attempts have been made to obtain a description of electron motion when inelastic collisions are significant. The simplest model for this situation is that due to Townsend (1925), for which it is assumed at each collision a constant fraction of the particle energy is lost. This fraction κ is assumed to be in excess of the value $2m/(M+m)$ appropriate to isotropic elastic collisions. Under these circumstances it is possible to set up an equation for the conservation of energy

$$\frac{d}{dt}\left(\frac{\overline{mv^2}}{2}\right) = eEv_D - \kappa\nu\left(\frac{\overline{mv^2}}{2}\right),\tag{1.3}$$

so that in equilibrium

$$\frac{\overline{mv^2}}{2} = \frac{eEv_D}{\kappa\nu} = \frac{e^2E^2}{\kappa m\nu},$$

or setting

$$k_B T_e = \tfrac{1}{2}\overline{mv^2}, \qquad k_B T_e = \frac{e^2E^2}{\kappa m\nu},$$

and

$$v_D^2 = \frac{\kappa}{2}\overline{v^2}.$$

Thus by the device of a field- or energy-dependent κ it is possible to go into the region where constant mobility is a reasonable approximation, while allowing there to be inelastic collisions. In this model then the drift motion in the field direction is determined by momentum-destroying collisions, while the field-dependent equilibrium temperature is given from the energy balance. This model is necessarily restricted to situations in which the excitation losses dominate over those leading to ionization.

The fractional energy loss quoted above is the average fraction of its energy an electron of a particular speed loses. In the gas kinetic situation it is appropriate to average over velocity distributions and this has been done by Cravath (1930), giving for electrons with a temperature T_e in a gas of temperature T_g,

$$\kappa = \frac{8}{3}\frac{mM}{(m+M)^2}\left(1 - \frac{T_g}{T_e}\right).$$

Fig. 1.5 shows measured values of κ for a range of gases indicating the qualitative differences between molecular and atomic gases and with low mean energy values of $8m/3M$ corresponding to $m \ll M$ and $T_e \gg T_g$. In general κ is dependent on the electron energy-distribution function.

An inelastic process which does not involve excitation is that of *charge transfer*. For ions it can be described by

$$X^+(\text{fast}) + Y(\text{slow}) \rightarrow X(\text{fast}) + Y^+(\text{slow}).$$

When X and Y are identical the probability is a maximum at zero relative energy; otherwise the probability has a maximum given by the uncertainty principle and the energy difference between the two states.

Under a wide range of conditions charge-exchanging collisions are the

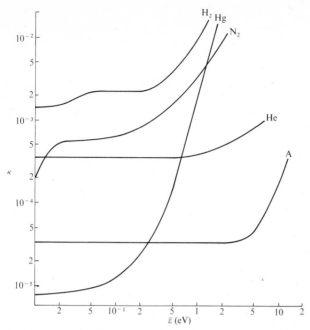

FIG. 1.5. Derived values for the fractional energy loss κ, per collision for electrons of mean energy $\bar{\varepsilon}$ in different gases. The theoretical value of $(8m/3M)$ for elastic collisions is seen to be appropriate at low energy and the onset of inelastic processes differs markedly in atomic and molecular gases.

dominant process for ions in their own gas, while for electrons they do not assume such importance.

1.3. Motion of charged particles in a magnetic field

In this situation in which a charged particle is acted on by a magnetic field *in vacuo* the equation of motion is

$$m \frac{d\mathbf{v}}{dt} = e\mathbf{v} \times \mathbf{B}, \qquad (1.4)$$

which can immediately be recognized as the equation for uniform circular motion with angular frequency $\omega_c \equiv eB/m$, the cyclotron frequency about **B**. The radius of the circle (Larmor radius) is determined by the transverse (perpendicular to **B**) velocity of the particle v_T and is given by $\rho = v_T/\omega_c$. Particles are 'tied' to the magnetic field lines.

The imposition of an electric field at right angles to the magnetic field causes a drift $\mathbf{E} \times \mathbf{B}/B^2$ perpendicular to both, as in Fig. 1.6. This can most readily be shown by introducing a velocity

$$\mathbf{v}' = \mathbf{v} - (\mathbf{E} \times \mathbf{B})/B^2;$$

Fɪɢ. 1.6. A typical path of an electron in crossed electric and magnetic fields *in vacuo*, showing drift perpendicular to both.

then we have

$$m\frac{d\mathbf{v}'}{dt} = e\left\{\mathbf{v}' \times \mathbf{B} + \mathbf{E} + \frac{(\mathbf{E} \times \mathbf{B}) \times \mathbf{B}}{B^2}\right\},$$

which, expanding the vector triple product and using $\mathbf{E} \cdot \mathbf{B} = 0$, gives

$$m\frac{d\mathbf{v}'}{dt} = e(\mathbf{v}' \times \mathbf{B}).$$

Hence the motion described by $\mathbf{v}$ is uniform circular motion with frequency ω_c superimposed on the steady drift of magnitude E/B perpendicular to E and B.

Collisions with other particles are required, however, in the uniform-field situation if there is to be any transverse mobility, i.e. motion in the direction of the electric field. Under these circumstances

$$m\frac{d\mathbf{v}}{dt} + m\nu\mathbf{v} = e(\mathbf{E} + \mathbf{v} \times \mathbf{B}) \tag{1.5}$$

and v_T, the component of $\mathbf{v}$ in the direction of $\mathbf{E}$ perpendicular to $\mathbf{B}$, is given by $v_T = \mu_T E$, with

$$\mu_T = \frac{\mu_0}{1 + \omega_c^2/\nu^2},$$

where μ_0 is the mobility in the absence of a magnetic field, previously written μ. This can most conveniently be shown by taking the vector product of (1.5) with $\mathbf{B}$ to give

$$m\nu(\mathbf{v} \times \mathbf{B}) = e\{\mathbf{E} \times \mathbf{B} + (\mathbf{v} \cdot \mathbf{B})\mathbf{B} - B^2\mathbf{v}\}, \tag{1.6}$$

assuming collisions to dominate ($\nu > \omega_c$) so that we may set $d\mathbf{v}/dt = 0$. Thus treating only the averaged motion we eliminate $\mathbf{v} \times \mathbf{B}$ between (1.5) and (1.6) to give

$$\left(\frac{m^2\nu^2}{e} + eB^2\right)\mathbf{v} = e\{\mathbf{E} \times \mathbf{B} + (\mathbf{v} \cdot \mathbf{B})\mathbf{B}\} + m\nu\mathbf{E},$$

and taking the scalar product with $\mathbf{E}$, and using $\mathbf{E} \cdot \mathbf{B} = 0$, gives the required result. Fig. 1.7 gives a diagram of such motion, where it can be

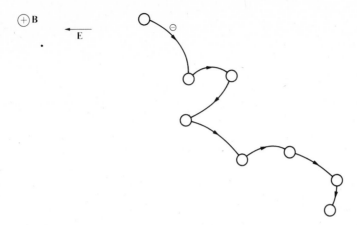

F𝕀G. 1.7. A typical path of an electron undergoing collisions in crossed electric and magnetic fields showing drift in the field direction and perpendicular to it.

seen that there is time-averaged drift motion also in the direction $\mathbf{E} \times \mathbf{B}$ of magnitude $(E/B) \cdot 1/(1 + v^2/\omega_c^2)$. Since the mobility is, on a binary or two-particle collision model, proportional to pressure, this implies that the effect of a magnetic field on charged particle motion is in some sense equivalent to an increase in gas pressure.

When a charged particle moves under the influence of both an alternating electric field and a steady magnetic field while undergoing collisions, its equation of motion is

$$m\frac{d\mathbf{v}}{dt} + m\nu\mathbf{v} = e(\mathbf{E} + \mathbf{v} \times \mathbf{B}).$$

$$\mathbf{E} = E_0\mathbf{i}\exp i\omega t, \qquad v = (v_x\mathbf{i} + v_y\mathbf{j})\exp i\omega t,$$

and thus

$$(i\omega + \nu)mv_x = eE_0 + ev_yB_0,$$

$$(i\omega + \nu)mv_y = -ev_xB_0,$$

giving

$$v_x = \frac{eE_0(i\omega + \nu)}{m\{\omega_c^2 + (\nu + i\omega)^2\}}, \qquad v_y = \frac{eE_0\omega_c}{m\{\omega_c^2 + (\nu + i\omega)^2\}}.$$

The power absorption is

$$\tfrac{1}{2}ne(\mathbf{v}^* \times \mathbf{E} + \mathbf{v} \times \mathbf{E}^*) \text{ per unit volume}$$

or

$$\frac{ne^2E_0^2\nu(\omega_c^2 + \nu^2 + \omega^2)}{(\omega_c^2 + \nu^2 - \omega^2)^2 + 4\nu^2\omega^2}, \tag{1.7}$$

which for $\nu \ll \omega$, ω_c is a resonance function with height at resonance (i.e. $\omega = \omega_c$) $\sim ne^2E_0^2/2\nu$ and half-width at half-height $\Delta\omega = \nu$.

Measurements of power-absorption conditions of cyclotron resonance

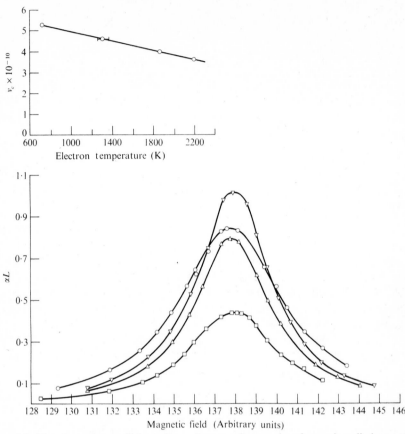

FIG. 1.8. The absorption coefficient αL at microwave frequencies of a caesium discharge of thickness L at a pressure of 0·0275 Torr and different temperatures showing a resonance at the electron cyclotron frequency as the magnetic field is varied. The collision frequency ν_e deduced from the absorption measurements is also shown. (Ingraham 1965.)

then should yield information on the collision frequency ν. Fig. 1.8 shows the power absorption and the variation of ν deduced from such a measurement at microwave frequencies in caesium vapour. Similar measurements at low temperature yield equivalent information on electrons in conductors and semiconductors (see Dresselhaus, Kip, and Kittel 1955).

1.4. Motion of an electron gas

So far the motion of single particles has been considered, and in the steady state this often suffices to give an accurate description of conditions in a gas discharge. However, the calculation and physical understanding of electric fields in the steady state and all variables in time-varying situations is aided by the introduction of two parameters, the

Debye length λ_D and the plasma frequency ω_p. The latter has a meaning even in the case where the particles do not have an effective temperature, while the former depends on the existence of random motion.

The simplest physical model which leads to there being a characteristic frequency is obtained by taking a finite slab containing equal numbers n of charged particles of both signs and displacing the electrons by an amount Δx, as in Fig. 1.9. This creates a surface charge causing there to be an internal field $E = en\,\Delta x/\varepsilon_0$ which causes the electrons to return to their original positions; their motion is given by

$$m\frac{d^2\,\Delta x}{dt^2} = -\frac{ne^2\,\Delta x}{\varepsilon_0},$$

and thus they oscillate with frequency ω_{pe}, where

$$\omega_{pe}^2 = \frac{ne^2}{m\varepsilon_0}. \tag{1.8}$$

The extension considering ion motion as well is straightforward, and because of the difference in mass between ions and electrons has little influence since the plasma frequency becomes

$$\omega_p = \left\{\frac{ne^2}{m\varepsilon_0}\left(1+\frac{m}{M}\right)\right\}^{\frac{1}{2}}.$$

The motion is collective, all particles having the same frequency.

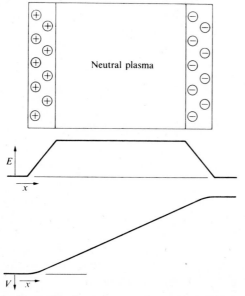

FIG. 1.9. A model of a plasma slab in which the ions and electrons have undergone a relative displacement Δx and the corresponding field E and potential V.

It is possible, therefore, to regard a region containing equal numbers of charged particles as a medium and to define a 'relative permittivity' for oscillating fields within the medium. When there are internal fields the electrons will undergo oscillations of amplitude x_0 given by $eE/m\omega^2$ and thus it shows a volume polarizability $P = -nex_0$; and since the electric displacement D is given by

$$D = \varepsilon_0 E + P \equiv \varepsilon_0 E \left(1 - \frac{ne^2}{\varepsilon_0 m \omega^2} \right),$$

the plasma can be characterized by a relative permittivity

$$\varepsilon = \varepsilon_0 \left(1 - \frac{\omega_{pe}^2}{\omega^2} \right). \tag{1.9}$$

The field associated with each particle in the medium will influence the motion of those in its immediate neighbourhood, and there will be a tendency for negative charges to be surrounded by positive charges and *vice versa*. Under appropriate conditions, namely in the solid state, this can lead to a regular array such as in the cubic lattice of ionic solids, e.g. sodium chloride. If the charged particles have sufficient energy, a regular spatially periodic arrangement cannot be maintained, and this is the case for ionic liquids and gases. There is thus a characteristic distance which determines whether the interaction between charged particles is essentially binary or many-body. If we treat the plasma as a medium with distributed charge, this distance will be given by comparing electrostatic energy and thermal energy, and using the model of uniformly displaced charge introduced to derive the plasma frequency, we equate the energy densities within the plasma

$$\frac{\varepsilon_0 E^2}{2} = \frac{nk_B T_e}{2},$$

regarding the electrons as a one-dimensional gas, to obtain a formula for the distance λ_D,

$$\frac{e^2 n^2 \lambda_D^2}{2\varepsilon_0} = \frac{nk_B T_e}{2} \quad \text{or} \quad \lambda_D^2 = \frac{\varepsilon_0 k_B T_e}{ne^2}. \tag{1.10}$$

The quantity λ_D is the Debye shielding distance and was first introduced in connection with ionic liquids. For distances short compared with λ_D a collective description of the plasma would not be expected to be valid and this is important in relation to wave motion for wavelengths λ shorter than λ_D. Also, in treating collisions between charged particles it is permissible to regard the Coulomb potential $e/\varepsilon_0 r$ as being screened and introduce a factor $\exp(-r/\lambda_D)$, provided that the number of particles within the distance λ_D is large, i.e. $n\lambda_D^3 \gg 1$.

1.5. Boltzmann's equation

In a wide variety of situations the rate of atomic processes depends strongly on the energy distribution of the interacting particles. This is particularly so when there is a threshold for some inelastic process. One therefore needs to find the energy distribution function in some self-consistent way. This means that a more detailed description of the collision processes is required.

A procedure which is convenient for description of motion in a steady uniform electric field is to define a distribution function $f(\mathbf{x}, \mathbf{v}, t)$ in terms of spherical harmonics about the field direction. A comprehensive treatment in this spirit has been given by Allis (1956). Then

$$f(\mathbf{x}, \mathbf{v}, t) = f_0(\mathbf{v}) + P_1(\cos \theta) f_1(\mathbf{v}) + P_2(\cos \theta) f_2(\mathbf{v}) + \dots,$$

where the Legendre functions $P_1(x) \equiv x$, $P_2(x) = \frac{1}{2}(3x^2 - 1)$. The collision integral which yields the term $(\partial f / \partial t)_c$ in the collisional Boltzmann equation

$$\frac{\partial f}{\partial t} + \mathbf{v} \cdot \frac{\partial f}{\partial \mathbf{x}} + \frac{e\mathbf{E}}{m} \frac{\partial f}{\partial \mathbf{v}} = \left(\frac{\partial f}{\partial t} \right)_c$$

is obtained by considering the details of trajectories before and after collision. These in general depend on the nature of the two particles, but approximations are available for the case in which the masses are disparate, e.g. electrons colliding with atoms. In this case, taking only elastic collisions into account, the spatially uniform steady state

$$\frac{\partial f}{\partial x} = \frac{\partial f}{\partial t} = 0$$

is found to be given by two equations corresponding to isotropic and $\cos \theta$ terms

$$\frac{1}{3v^2} \frac{d}{dv} \left(v^2 \frac{eE}{m} g_1 \right) = \frac{m}{Mv^2} \frac{d}{dv} \left\{ v^3 \nu \left(g_0 + \frac{k_B T_g}{m v} \frac{dg_0}{dv} \right) \right\}$$

and

$$\frac{eE}{m} \frac{dg_0}{dv} = -\nu g_1, \tag{1.11}$$

where g_0 and g_1 are isotropic distribution functions in terms of the speed v (see Allis 1956; Shkarofsky, Johnson, and Bachynski 1966). These can be combined to give

$$\frac{dg_0}{g_0} = -v \, dv \bigg/ \left(\frac{k_B T_g}{m} + \frac{M}{3m} \frac{e^2 E^2}{m^2 v^2} \right). \tag{1.11a}$$

Further integration is straightforward if ν is a known and well-behaved function of velocity. If it is taken as constant we recover a Maxwellian distribution with an effective electron temperature

$$T_{\text{eff}} = T_g + \frac{Me^2E^2}{3m^2\nu^2 k_B};$$

on the other hand, for a constant mean free path λ, $\nu = v/\lambda$ and we have

$$\frac{dg_0}{g_0} = -v\,dv \Big/ \left(\frac{k_B T_g}{m} + \frac{M}{3m}\frac{e^2 E^2 \lambda^2}{m^2 v^2}\right);$$

this integrates to,

$$\ln g_0 \propto -\frac{mv^2}{2k_B T_g} + \frac{ME^2\lambda^2 e^2}{6mk_B^2 T_g^2}\ln\left(v^2 + \frac{Me^2 E^2 \lambda^2}{3m^2 k_B T_g}\right),$$

which for sufficiently large E, namely,

$$E > \sqrt{\left(\frac{6m}{M}\right)}\frac{k_B T_g}{e\lambda},$$

can be shown by expanding the logarithmic term for small values of v to go over into the Druyvesteyn (1930) form for the bulk of the distribution,

$$\ln g_0 \propto -\frac{3}{4}\frac{m}{M}\frac{m^2 v^4}{e^2 E^2 \lambda^2}. \tag{1.12}$$

Typical figures suggest that for $E/p > 1 \cdot 0\ \mathrm{V\ m^{-1}\ Torr^{-1}}$ the departure from a Maxwellian at the gas temperature will be significant.

At high charged-particle densities it is necessary to extend the above by the simultaneous consideration of collisions between charged particles. Here a complication arises since the cross-section for binary encounters is infinite because the Coulomb inverse-square law of force is long-range, and the usual method of treatment is to introduce some cut-off or shielding length beyond which an encounter can no longer be regarded as binary, i.e. occurring between only two particles. To include such collisions one has to go back to the collision integral, when it is found that

$$\frac{1}{v^2}\frac{\partial}{\partial v}\left\{\left(\nu\frac{u_c}{3}v^3 + \alpha k_B T_e\right)\frac{\partial g_0}{mv\,\partial v} + \left(\frac{m}{M}\nu v^3 + \alpha\right)g_0\right\} = 0, \tag{1.13}$$

where

$$\alpha = \frac{\pi}{2}n\left(\frac{e^2}{2\pi\varepsilon_0 m}\right)^2 \ln\left\{\left(\frac{6\pi\varepsilon_0\lambda_D k_B T_e}{e^2}\right)^2 + 1\right\}\left\{\operatorname{erf}(x) - \frac{2}{\sqrt{\pi}}x\exp(-x^2)\right\},$$

with

$$x^2 = \frac{mv^2}{2k_B T_e},$$

so that

$$\alpha \simeq \sqrt{\left(\frac{\pi}{2}\right)}\left(\frac{e^2}{2\pi\varepsilon_0 m}\right)^2 \ln\left\{\left(\frac{6\pi\varepsilon_0\lambda_D k_B T_e}{e^2}\right)^2 + 1\right\}\frac{n}{3}\left(\frac{mv^2}{k_B T_e}\right)^{\frac{3}{2}}\left(1 - \frac{3mv^2}{10 k_B T_e}\right),$$

while

$$u_c = \frac{3mk_B T_g}{M} + \frac{e^2 E^2}{2mv^2} \equiv \frac{3m}{M}k_B T_{\text{eff}},$$

and it has been assumed that the electrons have an 'equilibrium' temperature T_e which may be different from T_g. Comparing the coefficients in (1.13) it can be seen that collisional effects involving charged particles dominate in the evolution of the distribution function for number densities,

$$n_e = \left(\frac{mk_B T_e}{2\pi}\right)^{\frac{1}{2}}\left(\frac{2\pi\varepsilon_0}{e^2}\right)^2 \frac{v}{N}\left(\frac{3mk_B T_g}{M} + \frac{e^2 E^2}{2mv^2}\right), \qquad (1.13a)$$

where

$$N = \tfrac{1}{2}\ln\{1 + (6\pi n_e \lambda_D^3)^2\}.$$

For small electric fields this can be rearranged into a physically more meaningful form by expressing in terms of the Debye length

$$\lambda_D = \left(\frac{\varepsilon_0 k_B T_e}{n_e e^2}\right)^{\frac{1}{2}}$$

and the plasma frequency

$$\omega_{pe} = \left(\frac{n_e e^2}{m\varepsilon_0}\right)^{\frac{1}{2}}$$

to read

$$\frac{v}{\omega_{pe}} < \frac{M}{m}\frac{T_e}{T_g}\frac{N}{6 \cdot 2^{\frac{1}{2}} \cdot \pi^{\frac{3}{2}}}\frac{1}{(n_e \lambda_D^3)},$$

while in the more usual field-dominated situation, the condition becomes

$$\frac{v}{\omega_{pe}} < 2^{\frac{1}{2}}\pi^{\frac{3}{2}}\frac{n_e \lambda_D^3}{N}\cdot\frac{eE\lambda_D}{k_B T_e}.$$

Under such conditions then the distribution function would be expected to approximate to a Maxwellian of temperature T_e, and while this was implicit in the assumptions made in writing down (1.13), criteria have been determined which establish limits on the expected observation of such a distribution.

Fig. 1.10 gives a comparison of Maxwellian and Druyvesteyn distributions having the same mean energy. In situations where excitation and ionization are occurring the important difference is the relatively fewer high-energy electrons, and it is sometimes convenient to use the Druyvesteyn form. This is discussed further in Appendix I. Figs 1.11 and 1.12

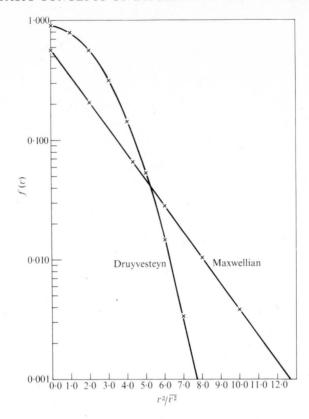

Fig. 1.10. A comparison of Maxwellian and Druyvesteyn distribution functions having the same mean-squared velocity.

give examples of actual distribution functions measured in the positive column in different gases under different conditions, showing quite close approximation in Fig. 1.11 to a Maxwellian distribution at low pressure (3 mTorr) in mercury, and in Fig. 1.12 a transition from Maxwellian to Druyvesteyn at a fixed point and different times τ in a discharge in neon with moving striations at quite high pressure (0·32 Torr). Both sets of data were obtained using the technique of double differentiation of the Langmuir-probe current–voltage characteristic to give the distribution function directly (Boyd and Twiddy 1959).

1.6. Plasma variables in terms of the distribution functions

It is usually convenient in situations where the charged particles have a distribution of velocities to introduce mean or fluid variables to describe

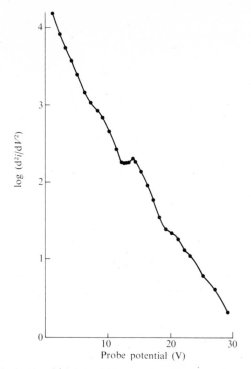

FIG. 1.11. The distribution function measured at low pressure (1·3 mTorr) in a mercury discharge showing a nearly linear $\log(\mathrm{d}^2 i/\mathrm{d}V^2)$ versus V plot, where i is the probe current and V its potential, corresponding to a Maxwellian distribution. (Rayment and Twiddy 1968.)

the average motion. These are usually the density n defined by

$$n_\alpha(\mathbf{x}, t) = \int_{-\infty}^{\infty} f_\alpha(\mathbf{x}, \mathbf{v}, t)\, \mathrm{d}\mathbf{v},$$

where the subscript α indicates that there may be several species of charged particle and $\int \mathrm{d}\mathbf{v}$ indicates a triple integral $\iiint \mathrm{d}v_x\, \mathrm{d}v_y\, \mathrm{d}v_z$, the drift velocity $\bar{v}_\alpha(\mathbf{x}, t)$ defined by

$$n_\alpha(\mathbf{x}, t)\bar{\mathbf{v}}_\alpha(\mathbf{x}, t) = \int_{-\infty}^{\infty} \mathbf{v}_\alpha(\mathbf{x}, t) f_\alpha(\mathbf{x}, \mathbf{v}, t)\, \mathrm{d}\mathbf{v},$$

and the mean energy $U_\alpha(\mathbf{x}, t)$ defined by

$$n_\alpha(\mathbf{x}, t) U_\alpha(\mathbf{x}, t) = \int_{-\infty}^{\infty} \tfrac{1}{2} m v_\alpha^2 f_\alpha(\mathbf{x}, \mathbf{v}, t)\, \mathrm{d}\mathbf{v}.$$

These terms are successive moments of the distribution function and by taking corresponding moments of the appropriate form of the Boltzmann equation one can deduce conservation equations.

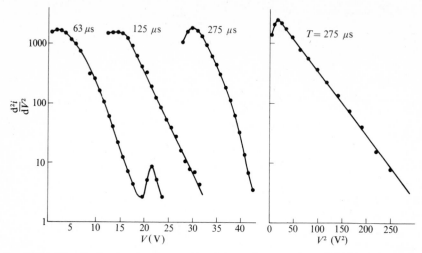

FIG. 1.12. The distribution function measured at medium pressures (0·32 Torr) in a neon discharge with moving striations of period 325 μs. It is seen early in the striation period (63 μs) to have a group of fast electrons which relax to give a closely Maxwellian distribution after 125 μs and this in turn develops into the Druvesteyn form late in the period (275 μs). (Rayment and Twiddy 1969.)

If in the course of collisions the identity of the particles does not change then the first moment is the same in both collisionless and collisional situations, and integrating the Boltzmann equation

$$\frac{\partial f_\alpha}{\partial t} + \mathbf{v}_\alpha \cdot \frac{\partial f_\alpha}{\partial \mathbf{x}} + \frac{e_\alpha \mathbf{E}}{m_\alpha} \cdot \frac{\partial f_\alpha}{\partial \mathbf{v}_\alpha} = \frac{\partial f_\alpha}{\partial t}\bigg|_{\text{coll}}.$$

gives

$$\frac{\partial n_\alpha}{\partial t} + \nabla \cdot (n_\alpha \bar{\mathbf{v}}_\alpha) = 0, \tag{1.14}$$

the equation of continuity or conservation of particles.

If, on the other hand, particles of the same type are created in the course of a binary collision as is the case for electrons when ionization by electron impact is occurring, the equation becomes

$$\frac{\partial n_\alpha}{\partial t} + \nabla \cdot (n_\alpha \bar{\mathbf{v}}_\alpha) = Z n_\alpha, \tag{1.14a}$$

where Z is the ionization rate.

Having introduced the drift velocity it is convenient to describe the particle motion by drift and random components, defining

$$\mathbf{v}_\alpha = \bar{\mathbf{v}}_\alpha + \mathbf{c}_\alpha,$$

where

$$\int \mathbf{c}_\alpha f_\alpha(\mathbf{x}, \mathbf{v}, t) \, d\mathbf{v}_\alpha = 0.$$

The second moment gives

$$\frac{\partial}{\partial t}(n_\alpha \bar{\mathbf{v}}_\alpha) + \nabla \cdot (\overline{n_\alpha \mathbf{v}_\alpha \mathbf{v}_\alpha}) + \frac{e_\alpha n_\alpha \mathbf{E}}{m_\alpha} = \int \mathbf{v}_\alpha \left. \frac{\partial f_\alpha}{\partial t} \right|_{\text{coll}} d\mathbf{v}_\alpha,$$

using the drift motion and the continuity equation (1.14) the left-hand side becomes

$$n_\alpha \frac{\partial \bar{\mathbf{v}}_\alpha}{\partial t} + n_\alpha \bar{\mathbf{v}}_\alpha \cdot \nabla(\bar{\mathbf{v}}_\alpha) + \nabla(\overline{n_\alpha \mathbf{c}_\alpha \mathbf{c}_\alpha}) + \frac{e_\alpha n_\alpha \mathbf{E}}{m_\alpha}, \tag{1.15}$$

where a bar over a quantity indicates an integration over the normalized distribution function and the quantities $\mathbf{v}_\alpha \mathbf{v}_\alpha$ and $\mathbf{c}_\alpha \mathbf{c}_\alpha$ are dyadics. The right-hand side can be written in the form $-\nu_\alpha n_\alpha (\bar{\mathbf{v}}_\alpha - \bar{\mathbf{v}})$, where the parameter ν_α is a frequency and, because of its appearance in what can be recognized as the equation of momentum for the fluid, is usually referred to as the collision frequency for momentum transfer, while $\bar{\mathbf{v}}$ is the mean fluid velocity

$$\frac{\sum m_\alpha \bar{\mathbf{v}}_\alpha}{\sum m_\alpha}.$$

ν_α is, of course, a lumped parameter and ignores the details of the collision process and velocity dependences, although these latter can be included where necessary. When under appropriate circumstances the third term of (1.15) is isotropic, it may be identified with the particle pressure and written $\nabla(n_\alpha k_B T_\alpha)$ by introducing the temperature T_α.

The third moment is concerned with energy and describes heat transfer. It is usual in gas-discharge situations to ignore the thermal effects and thus set the term $\nabla \cdot (n_\alpha \mathbf{v}_\alpha \mathbf{v}_\alpha \mathbf{v}_\alpha)$ identically zero. Once again introducing T_α one finds after manipulation

$$\frac{3}{2} \frac{\partial}{\partial t}(n_\alpha k_B T_\alpha) + \tfrac{5}{2} \nabla \cdot (n_\alpha k_B T_\alpha \bar{\mathbf{v}}_\alpha)$$

$$= \frac{e_\alpha n_\alpha \mathbf{E} \cdot \bar{\mathbf{v}}_\alpha}{m} - \kappa_\alpha \nu_\alpha n_\alpha k_B T_\alpha - \frac{\partial}{\partial t}\left(\frac{m_\alpha n_\alpha \bar{v}_\alpha^2}{2}\right), \tag{1.16}$$

where the second term on the right-hand side has a coefficient κ_α to indicate the mean fractional random energy loss at each collision occurring with a frequency ν_α. In the succeeding chapters we will be using the results of the somewhat restrictive assumptions made above; for a more general discussion, the reader is referred to Shkarofsky, Johnson, and Bachynski (1966). Some results for energy-dependent collision cross-sections are given in Appendix I. It should be noted that the results of § 1.2 follow directly from the uniform steady-state solution of the fluid momentum and energy equations, namely (1.15) and (1.16).

2

THE POSITIVE COLUMN

2.1. The self-sustaining glow discharge

THE regions of an active discharge in gases at low pressure were extensively studied and characterized at an early stage in the development of physics. Fig. 2.1 shows for a discharge at 1 Torr in a rare gas the approximate extent of the various regions. Basically, there must be a source of electrons or cathode, be it a cold metal plate, which emits secondary electrons because of ion bombardment, impact of metastable gas atoms, or sufficiently energetic ultraviolet radiation, or a hot thermionically emitting surface. Conditions adjacent to the cathode must be such as to give a sufficient yield of electrons, and in the case of ion bombardment this determines the electrical potential drop immediately in front of the cathode. Strong fields at the cathode result in the electrons being accelerated to energies greater than the optimum for causing ionization and there are, therefore, regions in which the initial electrons are scattered and lose energy until the motion is randomized and an optimum situation is achieved. When this is reached it is possible for there to be a balance between the energy input from the longitudinal field and the loss of energy due to inelastic collisions with the gas atoms, between the momentum gained from the electric field and that lost in elastic and inelastic collisions, and finally between the particle-generation process of ionization of the gas atoms by electron impact and the loss process of particle motion to the walls. The loss process arises because there are radial electric fields caused by the charged particles which lead to their motion in a radial direction and may be treated as free flight at very low pressures or diffusive at moderate pressures, particles of opposite sign then recombining on the wall. At high pressures when the reduced fields are low, recombination in the volume is an important process.

This equilibrium situation then can be axially uniform, and it is such a region which is referred to as the positive column of an active discharge. Because the number density of charged particles is maintained by ionization, their motion must be, by some criterion, collisional, but the importance of collisions will depend on the phenomena to be investigated. Since ionization from the ground state requires more energy transfer than any excitation process, there will be an appreciable density of excited

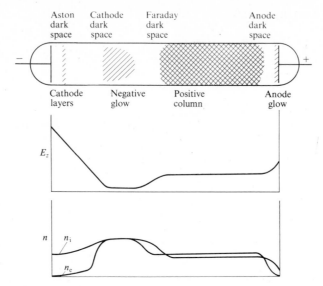

FIG. 2.1. Schematic diagram of a cold-cathode glow discharge at ~ 1 Torr indicating the regions with specific names associated. Also given are the longitudinal field E_z and the number densities n_i and n_e of positive ions and electrons. It is seen that only in the positive column is there an extended quasi-neutral region.

particles and consequently radiation will be emitted corresponding to allowed transitions between excited states of the gas atom, the spectra being characteristic of the gas used. The fact that the charged particles are generated in the volume and are lost at the walls means that there must be radial variations in the charged-particle density, so that the discharge is not radially uniform. On the other hand, if the radial fields are not so strong as to cause charge separation throughout the volume or, equivalently (as we shall see later), if the discharge is many Debye lengths in radius, the densities of positively and negatively charged particles will be almost equal.

Provided that the voltage source supplying the discharge is sufficient there is no limit to the longitudinal or axial extent of the positive column and so it is possible to set up experiments of any desired length.

Thus with the limitations noted above plasmas of appropriate characteristics can be generated. Let us determine some of these parameters. Typical cross-sections for ionization by electron impact are of the order of the classical cross-sectional area of an atom q_a, i.e. $10^{-20} \, \text{m}^2$, and thus the ionization rate per electron $Z = n_a q_a \bar{c}_e$, where n_a is the density of gas atoms and $\bar{c}_e$ the mean thermal speed of an electron, which at a pressure of 1 mTorr and an electron temperature of 1 eV gives a typical Z of $2 \times 10^5 \, \text{s}^{-1}$.

2.2. Calculation of ionization rates

Since the probability of ionization or excitation of an atom or molecule by electron impact is dependent on the electron energy, and in general in gas discharges the electrons have a distribution of energies, the rate of the process has to be determined by integration over the range of electron energies.

The incident electron must have an energy ε greater than the threshold ε_T for the process, and if the cross-section that the particular target gas atom presents is $Q(\varepsilon)$, then for an electron distribution function $g(\varepsilon)$ the rate Z is given by

$$Z = \int_{\varepsilon_T}^{\infty} g(\varepsilon) Q(\varepsilon) \left(\frac{2\varepsilon}{m}\right)^{\frac{1}{2}} d\varepsilon. \tag{2.1}$$

Equally, it can be determined from functions defined in terms of the electron velocity

$$Z = \int_{v_T}^{\infty} g(v) Q(v) v \, dv. \tag{2.1a}$$

The threshold for ionization is the ionization potential V_i, and for electron impact a close approximation to the cross-section is often a linearly increasing function of energy above V_i, say $Q = a_0 p(\varepsilon - \varepsilon_i)$, where a_0 is a constant and p is the gas pressure. Then for a Maxwellian distribution

$$\frac{Z}{p} = \int_{\varepsilon_i}^{\infty} \frac{2}{\sqrt{\pi}} a_0 (\varepsilon - \varepsilon_i) \left(\frac{2\varepsilon}{m}\right)^{\frac{1}{2}} \left(\frac{\varepsilon}{k_B T_e}\right)^{\frac{1}{2}} \exp(-\varepsilon/k_B T_e) \frac{d\varepsilon}{k_B T_e}$$

$$= \frac{4 a_0}{(2\pi m)^{\frac{1}{2}}} (k_B T_e)^{\frac{3}{2}} \left(2 + \frac{eV_i}{k_B T_e}\right) \exp\left(-\frac{eV_i}{k_B T_e}\right). \tag{2.2}$$

The maximum cross-section is typically at $3 - 4V_i$, and $eV_i \sim 2 - 10k_B T_e$, so the error in approximating Q by a linearly increasing function is small. Important differences occur for alkali metal vapours (see von Engel 1965).

At low pressures the drift velocity of electrons v_{De} is significant, and under these circumstances it is necessary to include this in any calculation of the ionization rate. The modified distribution function, which for a Maxwellian with superimposed drift will be proportional to

$$\exp\left[-\frac{m}{2k_B T_e} \{(v_x - v_{De})^2 + v_y^2 + v_z^2\}\right]$$

must then be used and $Z = Z(T_e, v_{De}/\bar{c}_e)$. Substitution in (2.1) followed by some manipulation gives the explicit expression

$$Z = \frac{a_0}{(2\pi m \Gamma_D)^{\frac{1}{2}}} \int_{\varepsilon_i}^{\infty} 2(\varepsilon - \varepsilon_i)\varepsilon^{\frac{1}{2}} \exp - \left(\frac{\varepsilon}{\varepsilon_m} + \Gamma_D\right) \text{sh} 2 \left(\frac{\Gamma_D \varepsilon}{\varepsilon_m}\right)^{\frac{1}{2}} d\varepsilon, \quad (2.2a)$$

where $\Gamma_D \equiv \varepsilon_D/\varepsilon_m = v_{De}^2/\overline{c_e^2}$, with $\varepsilon_D = \frac{1}{2}mv_{De}^2$ and $\varepsilon_m = \frac{1}{2}m\overline{c_e^2}$.

In general it is desirable to use the measured form for the ionization cross-section rather than the straight-line approximation which leads to (2.2) and (2.2a); but this implies numerical evaluation of Z and the relatively simple analytical dependence on the plasma variables is lost. It should be added that in normal positive columns only singly charged ions occur in significant numbers.

A method of measuring ionization rates is described in § 5.12.

2.3. The plasma balance equation

The volume generation rate must be balanced by a wall loss rate, and we shall see later that the effective wall-directed velocity v_w at low pressures is given approximately by $M_i v_w^2 = k_B T_e$, so that there must be a relation between generation and loss which for uniform radial density in a discharge of radius r_w would be $2v_w r_w = Zr_w^2$, or

$$Zr_w = 2\left(\frac{k_B T_e}{M_i}\right)^{\frac{1}{2}}. \quad (2.3)$$

In fact Q_a is a function of electron velocity and since Z is proportional to the neutral gas pressure p it is convenient to write eqn (2.3) in the form

$$\frac{Z}{p}(T_e) \cdot T_e^{-\frac{1}{2}} \cdot pr_w = \text{constant} \quad (2.4)$$

for a particular gas. This condition is charged-particle-density independent and shows that for a particular gas we would expect T_e to be a function of the group pr_w, which is known as a *similarity* variable of the discharge.

There are other such variables. Consider the mobility equation

$$mv_D v_e = eE.$$

The collision frequency for momentum transfer is proportional to density (or, at fixed gas temperature, to pressure) so that v_e/p is a constant for a particular gas; thus if the mobility relation is obeyed E/p must also be a similarity variable, since v_D is a function of T_e. From Poisson's equation for the space charge one can deduce directly that n_e/p^2 is a similarity variable. The current $I = n_e v_D \pi r_w^2 = (n_e/p^2)v_D \cdot \pi p^2 r_w^2$ is consequently also a similarity variable.

The criterion that the column radius should be large compared with the Debye length sets a lower limit to the charged-particle density n_0 if the positive column is to be considered a plasma, and since $\lambda_D^2 \equiv \varepsilon_0 k_B T_e / n_0 e^2$, $\lambda_D^2 / r_w^2 \equiv \varepsilon_0 k_B T_e / n_0 r_w^2 e^2$ shows $n_0 r_w^2$ to be a similarity variable. The criterion $r_w \gg \lambda_{De}$ for $r_w = 1$ cm, $T_e = 1$ eV is satisfied if $n_0 \gg 5 \cdot 3 \times 10^{11}$ m^{-3}.

An upper limit to the validity of the model outlined above is set by the assumption that charged-particle generation is by the single-stage process of direct impact. At higher charged-particle densities and at higher gas pressures, the processes of collisions between excited states and of electrons with excited states become important. To estimate appropriate parameters we need to calculate densities of excited states which will be determined by their generation and loss processes. For instance, a metastable state generated by electron impact and lost directly to the wall will have a density n_m given by

$$n_e n_a q_{am} \bar{c}_e = n_m \bar{c}_a / r_w,$$

where $\bar{c}_a$ is the atom mean thermal velocity and q_{am} is the cross-section for excitation from the ground state to the metastable state. Ionization by electron impact with metastable atoms then proceeds at a rate $n_m q_{mi} \bar{c}_e$, which has to be compared with $n_a q_{ai} \bar{c}_e$, where q_{ai} and q_{mi} are cross-sections for ionization from the ground state and metastable state respectively. The ratio will be greater than unity for

$$n_e > \frac{q_{ai} \bar{c}_a}{q_{am} q_{mi} \bar{c}_e r_w}.$$

Appropriate values for a mercury discharge, again for $r_w = 1$ cm, $T_e = 1$ eV, produce a value of $n_e \sim 10^{19}$ m^{-3}.

2.4. Collisions with boundary walls

So far it has been assumed that the collisions which charged particles undergo are with atoms or molecules. At low pressures, however, the mean free path or average distance between collisions may become comparable with the size of the vessel in which they move. Under these circumstances the effect of walls must be taken into account.

The detailed nature of processes at the wall will depend on its composition, temperature, and physical state generally. However, if we suppose that particle energies are not sufficient to cause secondary emission, (i.e. the creation of other charged particles), and that the wall is insulated and becomes negatively charged due to the greater mobility of electrons, some general conclusions can be made.

So far as electrons are concerned few reach the wall, since they are repelled by the adverse field, and therefore a reasonable model is one of specular reflection effectively at the wall with the momentum in the direction of the wall being reversed. For an assembly of electrons with mean thermal speed $\bar{c}_e$ and a lateral dimension L there would be an effective collision frequency $\bar{c}_e/L$.

Ions, on the other hand, will be attracted to the wall and neutralized there. In a symmetrical situation the maximum path length normal to the wall will be $L/2$ and the time taken will depend on the flight time of the ion in the accelerating field. If its mean velocity is normal to the boundary $\bar{v}_n$ then there will be an effective collision frequency $2\bar{v}_n/L$. Direct experimental measurements to test the models above are difficult, but there is much indirect evidence. Fig. 2.2 shows measurements of the effective collision frequency for electrons at microwave frequencies in a low-pressure mercury discharge, the theory for which will be given later. It can clearly be seen that as the pressure is reduced there is a transition from $\nu_{eff} \propto p$, corresponding to collisions with gas atoms, to $\nu_{eff} \propto \bar{c}_e$, corresponding to collisions with the walls. The simplest model for the transition is to add the collision frequencies $\nu_{total} = \nu_{walls} + \nu_{gas}$.

We will need to discuss the influence of collisions with the walls on ion motion later in connection with ion waves (Chapter 7).

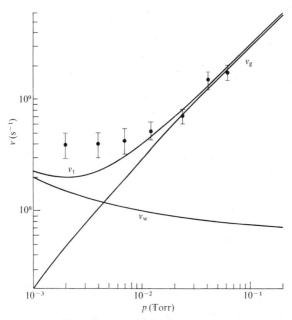

FIG. 2.2. Measurements of the effective collision frequency ν_t for electrons at low pressure in mercury compared with the theoretical curves for collisions with the walls ν_w and gas atoms ν_g and $\nu_t = \nu_g + \nu_w$. (Bryant and Franklin 1963.)

2.5. A simple model of a plane positive column

If the following assumptions are made a simple model, but one with a number of the essential features of the positive column at low and medium pressures and normal current densities, is obtained:

(1) the discharge x, or lateral, dimension L is large compared with the Debye length λ_D, so that space-charge fields are small and the approximation of *quasi-neutrality* $n_e = n_i$ can be made;

(2) ionization occurs as a single-stage process by electron impact on gas atoms;

(3) ion motion is determined by ion inertia, collisions, and electric fields, while pressure effects are negligible, which is tantamount to setting the ion temperature $T_i = 0$;

(4) electron motion is determined by collisions, pressure effects, and electric fields, while because of the relatively light mass of an electron its inertia is negligible;

(5) the parameters do not vary in the longitudinal direction of the discharge;

(6) the electron temperature is assumed to be given by the longitudinal field and to be independent of x.

With these assumptions and setting $n_e = n_i = n$ in advance, the equations of continuity become

$$\frac{\mathrm{d}}{\mathrm{d}x}(nv_{ex}) = Zn, \qquad (2.5)$$

$$\frac{\mathrm{d}}{\mathrm{d}x}(nv_{ix}) = Zn, \qquad (2.6)$$

from which with the condition $v_{ex} = v_{ix} = 0$ at the centre $x = 0$, one can deduce $v_{ex} = v_{ix} = v(x)$. A schematic diagram of the variables is given in Fig. 2.3.

The momentum equation for ions becomes

$$Mv\frac{\mathrm{d}v}{\mathrm{d}x} + Mv\nu_i + MvZ = eE, \qquad (2.7)$$

using the fact that

$$\nabla \cdot (n\mathbf{vv}) \equiv \mathbf{v}\nabla \cdot (n\mathbf{v}) + (n\mathbf{v}\cdot\nabla)\mathbf{v}$$

and that for electrons

$$mn\nu_e v + mnZv = -\frac{\mathrm{d}}{\mathrm{d}x}(nk_B T_e) - neE. \qquad (2.8)$$

The three variables n, v, and E can be found as functions of x by solving subject to the appropriate boundary conditions. It is convenient, and

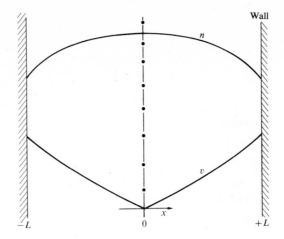

FIG. 2.3. Cross-section of a plane positive column indicating the coordinate system to be used and expected variation of charged particle density and transverse particle speed.

gives physical insight, to normalize the equations by introducing new variables defined as follows

$$\frac{d\eta}{dx} = \frac{eE}{k_B T_e}, \qquad u = \frac{v}{c_s},$$

where $c_s^2 = k_B T_e/M$,† $\xi = xZ/c_s$, and $N = n(\xi)/n(0)$, to give

$$\frac{d}{d\xi}(Nu) = N, \tag{2.9}$$

$$u\frac{du}{d\xi} + u\,\delta_i = \frac{d\eta}{d\xi}, \tag{2.10}$$

$$Nu\,\delta_e = -\frac{dN}{d\xi} - N\frac{d\eta}{d\xi}, \tag{2.11}$$

where $\delta_i = 1 + (\nu_i/Z)$ and $\delta_e = (m/M)((\nu_e + Z)/Z)$ are measures of the particle collision frequencies for momentum transfer relative to the ionization frequency. The ratio, $\delta_e/\delta_i = (m/M)((\nu_e + Z)/(\nu_i + Z))$, which for low values of T_e or $Z \ll \nu_i$ gives $\delta_e/\delta_i \sim \mu_i/\mu_e$, while for high values of T_e, $\delta_e/\delta_i \sim m/M$; it thus ranges from 10^{-2} to 10^{-5}. Introducing $\lambda_N \equiv \ln N$ the differential equations become first order in the variables u, λ_N, and η and the coefficients are functions solely of u. This means that it is most convenient

---

† We shall see in Chapter 7 that c_s is the isothermal ion sound speed the characteristic phase velocity of ion waves.

to solve them with u as the independent variable. This gives

$$\frac{d\xi}{du} = \frac{1-u^2}{1+u^2(\delta_e+\delta_i)}, \tag{2.12}$$

$$\frac{d\lambda_N}{du} = \frac{-u(1+\delta_e+\delta_i)}{1+u^2(\delta_e+\delta_i)}, \tag{2.13}$$

$$\frac{d\eta}{du} = \frac{u(1+\delta_i)+u^3\,\delta_e}{1+u^2(\delta_e+\delta_i)}. \tag{2.14}$$

The boundary conditions at $\xi = 0$ are $u = 0$ by definition, $\eta = 0$, since the zero potential has to be defined, and $u = 0$ on physical grounds of symmetry. The outer or wall boundary condition is set by the requirement that the variables u, λ_N, and η must be single-valued functions of ξ, and this will require further discussion. Writing $\delta_e + \delta_i = \delta$ gives some simplification, and one finds

$$\xi = -\frac{u}{\delta} + \frac{\delta+1}{\delta^{\frac{3}{2}}}\tan^{-1}(u\,\delta^{\frac{1}{2}}), \tag{2.15}$$

$$\lambda_N = -\frac{(1+\delta)}{2\,\delta}\ln(1+u^2\,\delta), \tag{2.16}$$

$$\eta = \frac{\delta_e u^2}{2\,\delta} + \frac{\delta_i(1+\delta)}{2\,\delta^2}\ln(1+u^2\,\delta). \tag{2.17}$$

The requirement that u is a single-valued function of ξ can now be discussed, and to do this we need to consider both (2.12) and (2.15). Since $u = 0$ at $\xi = 0$ and $du/d\xi \sim 1$ for small u, u increases monotonically with ξ until it reaches the value unity; at this point the gradients of number density and potential are infinite. The point is identified with the wall, although as yet we only prove it to be an upper limit for the wall position; and the boundary condition of the directed velocity v being equal to $(k_B T_e/M)^{\frac{1}{2}}$ we shall see later coincides with the Bohm (1949) criterion for the formation of a space-charge sheath.

It can readily be shown from eqn (2.15) that for large values of δ the proper variables for the problem are $\xi\delta^{\frac{1}{2}}$ and $u\delta^{\frac{1}{2}}$, whence one finds $u\delta^{\frac{1}{2}} \sim \tan \xi\delta^{\frac{1}{2}}$ and therefore $\xi_w\,\delta^{\frac{1}{2}} \leqslant \pi/2$. This coincides with the value given by the Schottky (1924) theory of the positive column which ignores inertial effects and is collision-dominated. On the other hand, as $\delta \to 1$ the equations become

$$\xi \sim u - \frac{2u^3}{3}, \qquad \lambda_N \sim -u^2, \qquad \eta \sim u^2,$$

with

$$\xi_w = \frac{\pi}{2} - 1, \qquad \lambda_N = -\eta_w = -\ln 2.$$

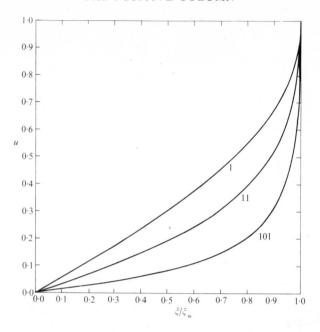

FIG. 2.4. Computed normalized variation of the transverse particle speed $u = v/c_s$ where $c_s = (k_B T_e/M)^{\frac{1}{2}}$ with normalized transverse position ξ for different values of the collision parameter δ for a plane discharge.

The manner in which number density and radial velocity vary with normalized radius is shown in Figs 2.4 and 2.5 for a range of values of δ.

2.6. The free-fall model of the positive column

It is of interest to compare the results of the above model with those obtained by Tonks and Langmuir (1929), who gave what is, under some assumptions at low pressure, a more exact treatment of the ion motion by treating the ions as being generated at rest and then falling freely to the walls. This results in there being a distribution of ion radial velocities at any point in the discharge. An ion which was generated at $x = x_0$ then has a speed at x_1 of

$$v_i(x_1, x_0) = [(2e/M)\{V(x_0) - V(x_1)\}]^{\frac{1}{2}};$$

setting $\delta_e = 0$ in (2.11) and normalizing, as in the previous section, the ion velocity and space coordinate, the ion density is found to be

$$\frac{1}{\sqrt{2}} \int_0^\xi \frac{\exp\{-\eta(\sigma)\}\,\mathrm{d}\sigma}{\{\eta(\xi) - \eta(\sigma)\}^{\frac{1}{2}}},$$

and therefore setting the electron and ion densities equal, η is given by

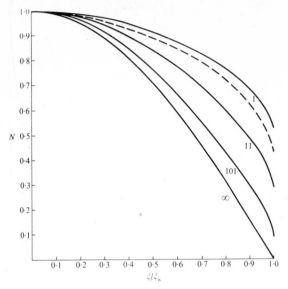

F<small>IG</small>. 2.5. Computed variation of the relative charged-particle density $N = n/n_0$ as a function of ξ with δ as a parameter. The corresponding free-fall distribution is shown dashed.

the solution of

$$\exp\{-\eta(\xi)\} = \frac{1}{\sqrt{2}} \int_0^\xi \frac{\exp\{-\eta(\sigma)\}\, d\sigma}{\{\eta(\xi) - \eta(\sigma)\}^{\frac{1}{2}}}. \tag{2.18}$$

The solution of this integral equation can be expressed in various ways, for instance, using the Dawson function (see Harrison and Thompson 1959), but was first given in the form of a power series. There is a singularity of similar nature to that found above in the fluid approximation at specific values of ξ and η, these being $\xi_w = 0.5722$, $\eta_w = 0.8539$, hence $N_w = 0.4258$. The variation of number density with normalized radius is compared with the fluid model in Fig. 2.4.

2.7. Extending the models

It was noted at the outset that certain assumptions were being made in setting up the model; these will be relaxed one by one and their consequences examined, but some will be deferred to later chapters.

First, let us consider the assumption of plane geometry. There is considerable experimental difficulty in setting up a stable discharge with anything approaching uniformity in plane geometry; the most common experimental situation is one in which the discharge is cylindrical. The extension to cylindrical geometry is straightforward, and when the divergence operator is written in terms of cylindrical coordinates with azimuthal symmetry $(\partial/\partial\theta \equiv 0)$ one finds that the continuity equation becomes

$$\frac{d}{d\xi}(\xi Nu) = N\xi, \tag{2.9a}$$

the ion motion

$$\frac{d}{d\xi}(\xi Nu^2) + \frac{\nu_i}{Z}\xi uN = N\xi\frac{d\eta}{d\xi}, \tag{2.10a}$$

and the electron motion

$$\frac{dN}{d\xi} = -N\frac{d\eta}{d\xi} - Nu\,\delta_e. \tag{2.11a}$$

This situation was first considered by Woods (1965) for ν_i and $\delta_e = 0$, and in general the solution can be found in a similar way to the plane case, but the integration of the differential equations must be performed numerically. Again it is convenient to use the radial velocity as the independent variable and the form of the equations, e.g.

$$\frac{du}{d\xi} = \frac{u^2(\delta_e + \delta_i) + 1 - u/\xi}{1 - u^2}, \tag{2.12a}$$

shows $u = 1$ to be a singular point. The boundary value of $\xi = \xi_w$ is a function of δ_i and δ_e, and the balance between generation and loss rate defines ξ_w as an eigenvalue of the problem. For $\delta_i = 1$, $\delta_e = 0$, ξ_w is found to be $1\cdot1094$, while as δ_i, $\delta_e \to \infty$ the solution has the form of the Schottky solution.

The variations of number density and radial velocity are shown for a range of values of δ in Figs 2.6 and 2.7. Comparison can be made with the corresponding ion free-fall results (Tonks–Langmuir) and the boundary values are compared in Table 2.1. The equation relating ξ_w to the plasma parameters is a particular case of the plasma balance relation (2.4). The free-fall model can readily be treated in cylindrical geometry, and the normalized number densities in plane and cylindrical discharges are shown in Fig. 2.8; characteristic values are given in Table 2.1 (Self 1965).

In the collision-dominated case the equations are more appropriately expressed in terms of different variables as mentioned in connection with the plane case, and taking $\nu_i \gg Z$ gives

$$\frac{1}{r}\frac{d}{dr}(nvr) = Zn, \qquad Mv\nu_i = e\frac{dV}{dr}, \qquad mv\nu_e = -e\frac{dV}{dr} - \frac{k_B T_e}{n}\frac{dn}{dr}.$$

Eliminating first V and then nv gives

$$\frac{d^2n}{dr^2} + \frac{1}{r}\frac{dn}{dr} + \frac{Zn}{D_a} = 0, \tag{2.19}$$

where D_a is the ambipolar diffusion coefficient

$$D_a \equiv \frac{k_B T_e}{M\nu_i + m\nu_e} = \frac{\mu_i\mu_e k_B T_e}{e(\mu_e + \mu_i)} \sim \frac{\mu_i k_B T_e}{e} = \frac{\mu_i}{\mu_e}\cdot D_e, \tag{2.20}$$

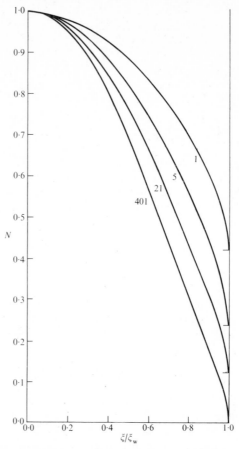

FIG. 2.6. The relative density $N = n/n_0$ as a function of the normalized radial coordinate for a cylindrical discharge with δ as a parameter.

and where D_e is the electron diffusion coefficient. The eigenvalue of this equation for $n \to 0$ as $r \to r_w$ or equivalently $dV/dr \to \infty$ is the first zero of the Bessel equation, and thus $r_w(Z/D_a)^{\frac{1}{2}} = 2 \cdot 4048$. Consider now the variable $\xi \delta^{\frac{1}{2}}$. At low pressures $\nu_i \to 0$ and $\delta_e \to 0$ so that $\delta \to 1$ and $\xi_w \delta^{\frac{1}{2}} = 1 \cdot 1094$ in cylindrical geometry. At high pressures

$$\xi_w \, \delta^{\frac{1}{2}} \equiv \frac{r_w Z}{c_s} \left(1 + \frac{\nu_i}{Z} + \frac{m}{M} \frac{\nu_e + Z}{Z} \right)^{\frac{1}{2}}$$

$$= \frac{r_w Z}{c_s} \left(1 + \frac{m}{M} + \frac{\nu_i}{Z} + \frac{m\nu_e}{MZ} \right)^{\frac{1}{2}}$$

$$\sim r_w \left(\frac{Z}{D_a} \right)^{\frac{1}{2}} \quad \text{for} \quad \nu_i \gg Z.$$

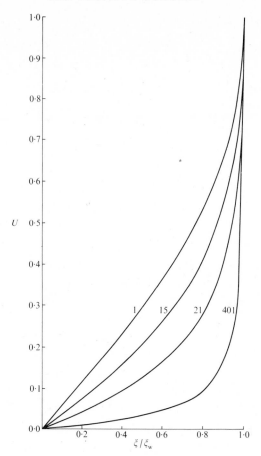

FIG. 2.7. Normalized particle speed u as a function of normalized radial coordinate with δ as a parameter.

TABLE 2.1

Characteristic values for collisionless discharges

	ξ_w	N_w	η_w	U_w
Plane fluid	0·5708	0·5000	0·6931	1·00
Plane free fall	0·5722	0·4258	0·8539	1·35
Cylindrical fluid	1·1094	0·4195	0·8687	1·00
Cylindrical free fall	1·0916	0·3150	1·1551	1·54

$$\xi_w = \frac{x_w Z M^{\frac{1}{2}}}{(k_B T_e)^{\frac{1}{2}}}, \qquad N_w = \frac{n_w}{n_0}, \qquad \eta_w = \frac{e V_w}{k_B T_e}, \qquad U_w = \frac{\overline{v^2} M}{k_B T_e}$$

are the normalized wall position, number density, potential, and mean ion energy respectively.

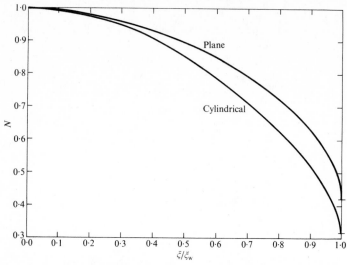

FIG. 2.8. A comparison of the transverse density distributions for the free-fall model with ionization by electron impact in plane and cylindrical geometries.

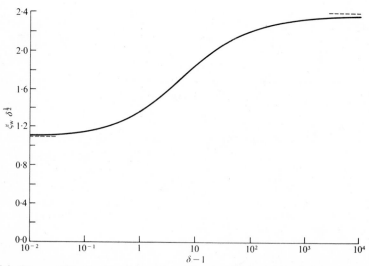

FIG. 2.9. The normalized wall position $\xi_w = r_w Z / c_s$ as a function of the collision parameter δ which is most conveniently displayed by plotting $\xi_w \delta^{\frac{1}{2}}$ versus $\delta - 1 \equiv \nu_i / Z$, showing asymptotic limits corresponding to the Schottky theory at large δ and to a value close to the free-fall theory for small ν_i.

Thus $\xi_w \delta^{\frac{1}{2}} \to 2\cdot4048$ for large δ and therefore $\xi \delta^{\frac{1}{2}}$ is a suitable variable with which to describe the transition of the plasma balance relation from collisionless to collisional plasmas. Fig. 2.9 shows its variation over six decades in δ.

The fact that the results for the fluid model give a smooth transition from agreement of better than $0\cdot5$ per cent with the free-fall model to identically the diffusion-dominated model and spans the region of parameters in which the majority of low pressure and afterglow plasmas is what gives the model its usefulness, and it is therefore natural to seek to extend and apply it further.

Now using the fact that the plasma balance relation is essentially an equation relating the electron temperature to the similarity variable pr_w, as indicated in the introduction to this chapter, we can obtain for any particular gas the curve for $T_e(pr_w)$ using the appropriate values for ν_i/p and ν_e/p if $(Z/p)(T_e)$ is known. In this way the transition from Langmuir–Tonks to Schottky theories can again be described, and Fig. 2.10 gives results for mercury obtained by Forrest and Franklin (1966).

Also, while maintaining charge neutrality, one can examine the consequences of T_i being non-zero. It is a relatively simple algebraic exercise to show that if this is done and the normalization of radial speed is made with respect to c_s now defined by $c_s^2 = k_B(T_e + T_i)/M$, then the problem is formally unaltered.

Further, one can seek to introduce different models for the creation of charged particles. The simplest and physically most difficult to justify is to replace the generation term Zn_e by Yn_e^2. The results of such a modification are shown in Fig. 2.11, where it can be seen that the physics of the situation is little altered. A more meaningful addition to the generation

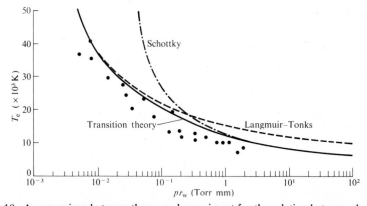

FIG. 2.10. A comparison between theory and experiment for the relation between electron temperature and the similarity variable pr_w for mercury. The experimental values are from Granovsky (1940), Killian (1930), and Klarfeld (1939). Drift is included in the theory and the high-and low-pressure theoretical models are shown dashed.

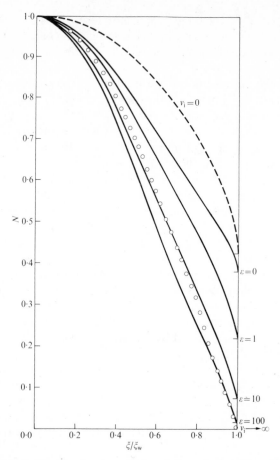

Fig. 2.11. Theoretical curves for the radial variation of charged-particle density in a 'discharge' for which ionization is entirely by processes such that the rate Z is proportional to electron density squared. The parameter ε is defined by $\varepsilon = \nu_i/Z_0$, Z_0 being the central generation rate per electron. For comparison curves are given for $Z \propto n$ at two limiting values of $\delta - 1$.

and loss processes would be a volume loss process—recombination—again proportional to n_e^2. In this case there can be significant changes in the charged-particle distribution, and we will return to this in a discussion of the contraction of the positive column later.

If one is to include the effects of two-stage ionization processes in anything like a reasonable physical model, the details of possible processes for a particular gas need to be known. This implies that the cross-sections for the most important excitation processes should be available, and that ionization cross-sections from these states should be known; the

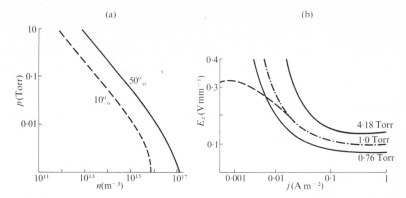

FIG. 2.12. Curves showing the importance of two-stage processes in mercury as a function of gas pressure p and electron number density n. (a) the fraction of ionizing processes due to two-stage processes; (b) the corresponding influence of the current density j on the longitudinal field E_z, taking into account also space-charge effects. The solid curves represent experimental measurement by Mierdel and Schmalenberg (1936). The dashed curve represents the effect of the transition from two-stage to single-stage processes alone. The chain curve includes space-charge effects as indicated in Chapter 4.

existence of metastable states also needs to be accounted for. Further it is important in such a situation to recognize that the electron distribution function cannot be described by a single parameter, the electron temperature. Equations taking into account the radial variations of densities and distribution functions become impossibly difficult to solve. It is worth listing some treatments in the literature in different approximations which have attempted to take these factors into account: Gentle (1966) in argon; Mewe (1967) in helium; Swain and Brown (1971) in argon; and Forrest and Franklin (1969) in mercury. Results of these latter authors estimate when two-stage processes become important and compare measurements of the longitudinal field as a function of current with theory; they are given in Fig. 2.12. To include the effect of space charge at low currents some of the results of Chapter 4 are needed.

3

MAGNETIC FIELD EFFECTS

3.1. The positive column in a magnetic field

FOR A number of reasons many experiments involving plasma effects in positive columns have been carried out with an axial magnetic field. There is the practical reason that discharges are often more stable and less noisy under such conditions and the scientific one that the wealth of interesting processes and effects is enhanced by the existence of a steady magnetic field.

The effect of an axial magnetic field is typically markedly different for electrons and for ions, and depends on the ratio of the appropriate cyclotron frequency to collision frequency. Since $\omega_{ce} = eB/m$ is equivalent to a frequency $f_{ce} = \omega_{ce}/2\pi$ of $28B$ MHz (B in telsa) and $\nu_e \sim 3 \times 10^9 \times p \, s^{-1}$ (p the gas pressure in Torr), $\omega_{ce} > \nu_e$ for $B/p > 0 \cdot 016 \, T \, Torr^{-1}$; under these circumstances the electrons can be described as magnetized. Similarly for ions, $\omega_{ci} = eB/M$ ($f_{ci} = 16B$ MHz for H^+, B in tesla) and ν_i is $\sim 5 \times 10^7 \times p \, s^{-1}$, $\omega_{ci} > \nu_i$ for $B/p > 0 \cdot 5 \, T \, Torr^{-1}$, and thus there is typically a range of two orders of magnitude of magnetic field or pressure in which the electrons are magnetized and the ions are not. Physically, an axial magnetic field inhibits lateral motion, be it diffusion or more nearly free fall, and since radial losses are reduced a discharge can be maintained with a smaller ionization rate—i.e. a lower electron temperature.

It is possible to set up a model of the positive column in the presence of a magnetic field along the lines of Chapter 2, but from the foregoing discussion it is apparent that there is a range of fields for which the ion motion may be substantially unaffected, while the electron motion will be significantly modified.

The electron momentum equation then is

$$(Z + \nu_e)\mathbf{v}_e = -e \frac{\mathbf{v}_e \times \mathbf{B}}{m} - \frac{e}{m} \mathbf{E} - \frac{k_B T_e}{m} \frac{\nabla n_e}{n_e}. \tag{3.1}$$

If we assume variation of n_e and $\mathbf{E}$ only in the x-direction then the equation becomes

$$\frac{k_B T_e}{m n_e} \frac{dn_e}{dx} = -\frac{eE}{m} - v_{ey} \frac{eB}{m} - (Z + \nu_e)v_{ex}, \tag{3.2}$$

and
$$0 = v_{ex}\frac{eB}{m} - (Z + \nu_e)v_{ey},$$ (3.3)

and the x motion can be found by substituting (3.3) in (3.2) to be given by

$$(Z + \nu_e)v_{ex}\left\{1 + \frac{\omega_{ce}^2}{(\nu_e + Z)^2}\right\} = -\frac{eE}{m} - \frac{k_B T_e}{mn_e}\frac{dn_e}{dx}.$$

The effective electron diffusion coefficient is thus

$$D_{em} = k_B T_e \left/ \left[m(Z + \nu_e)\left\{1 + \frac{\omega_{ce}^2}{(\nu_e + Z)^2}\right\}\right],\right.$$

and is thus reduced by a factor

$$\left\{1 + \frac{\omega_{ce}^2}{(\nu_e + Z)^2}\right\}\left(1 + \frac{Z}{\nu_e}\right)$$

in comparison with the magnetic field and ionization-free case. Customarily this is written in the form where ionization is ignored, i.e.

$$D_{em} = \frac{D_{e0}}{(1 + (\omega_{ce}^2/\nu_e^2))}$$

(see e.g. Chapman and Cowling 1939). Similar results hold in cylindrical geometry with r replacing x and θ replacing y. The electrons then have an azimuthal motion with

$$v_{e\theta} = v_{er}\frac{\omega_{ce}}{(\nu_e + Z)}.$$

In the same spirit as eqns (2.12)–(2.14), we derive equations for number density, radial velocity, and potential, retaining finite ion temperature. They are found to be, after some manipulation and using the normalizations $\eta = -eV/k_B T_e$; $N = n_e/n_{e0}$; $u = v_i/c_s$; $\xi = rZ/c_s$,

$$\frac{du}{d\xi} = \frac{1 - u/\xi + \delta u^2}{1 - u^2},$$ (3.4)

$$\frac{d\eta}{d\xi} = -\frac{u^2/\xi + u + \delta_i u + \delta_e u^3(1 + T_i/T_e) - \delta_e u T_i/T_e}{1 - u^2}$$ (3.5)

$$\frac{1}{N}\frac{dN}{d\xi} = \frac{u^2/\xi - u - \delta u}{1 - u^2},$$ (3.6)

and thus are very similar to those found in Chapter 2 in that there is a singularity once again where $u = 1$. However, δ_e is now

$$\delta_{em} \equiv \frac{k_B T_e}{MZD_{em}} \equiv \frac{m}{M}\left[1 + \frac{\omega_{ce}^2}{(\nu_e + Z)^2}\right]\left(1 + \frac{\nu_e}{Z}\right),$$ (3.7)

while for conditions in which the ions remain unmagnetized δ_i is still $(1 + \nu_i/Z)$, thus the ratio δ_i/δ_e is now no longer

$$\left(\frac{M\mu_e}{m\mu_i}\right)\frac{(1 + Z/\nu_i)}{(1 + Z/\nu_e)}$$

but is multiplied by a factor

$$\left\{1 + \frac{\omega_{ce}^2}{(\nu_e + Z)^2}\right\}^{-1}$$

which can cause the ratio to become less than unity. Thus the imposition of a magnetic field has the effect of increasing δ, i.e. effectively increasing the gas pressure or reducing the electron temperature. The results of Chapter 2 for n and v thus can be used directly with a modified δ. The effect on the radial field is a little more complicated and needs further examination. From (3.4) it can readily be shown that for small u, $u \sim \xi/2$, and this allows us to examine (3.5)

$$\frac{d\eta}{d\xi} \sim \xi\left(\frac{3}{4} + \frac{\nu_i}{2Z} - \frac{\delta_{em}T_i}{2T_e}\right).$$

Therefore for

$$\delta_{em} > \frac{T_e}{T_i}\left(\frac{\nu_i}{Z} + \frac{3}{2}\right)$$

the potential initially increases with ξ. However, where

$$u^2 \approx \frac{T_i}{(T_i + T_e)}$$

the sign of $d\eta/d\xi$ changes, and the behaviour near the wall is similar to that in the absence of a magnetic field. Results in Figs 3.1 and 3.2 show how, as δ_{em} increases, this field reversal manifests itself.

Comparison can be made with experimental results both in terms of distributed and bulk variables, and Fig. 3.3 gives the results of Bickerton and von Engel (1956) for the ratio of wall density to central density compared with theory as developed by Forrest and Franklin (1966a) and by Crawford, Ewald, and Self (1967). Fig. 3.4 gives a comparison of the electron temperature as measured by microwave noise emission by Forrest and Franklin (1967) with that deduced from the plasma balance equation

$$\xi_w(\delta_e, \delta_i) = pr_w\frac{Z}{p}(T_e)\left(\frac{M}{k_BT_e}\right)^{\frac{1}{2}}, \tag{3.8}$$

taking into account the variation of δ_e with magnetic field. The presence

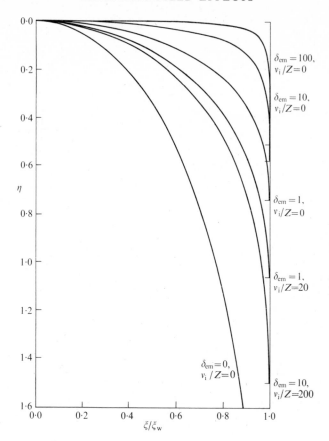

FIG. 3.1. The radial potential profile for different values of the parameters

$$\frac{\nu_i}{Z} \quad \text{and} \quad \delta_{em} \equiv \left\{ 1 + \frac{\omega_{ce}^2}{(\nu_e + Z)^2} \right\} \left(1 + \frac{\nu_e}{Z} \right) \cdot \frac{m}{M}$$

of a magnetic field in modifying the electron number-density distribution causes a change in light emission, and this 'contraction' of the column is displayed in Fig. 3.5, which gives measurements of Forrest and Franklin (1966b) compared with the fluid theory for appropriate values of δ. More gross effects of 'axial' magnetic fields have been found in situations where the fields are of limited extent and therefore are non-uniform and have significant radial components as well (see Rhoklin (1939) and in connection with ion wave excitation, Little and Jones (1965) and Barrett (1966)).

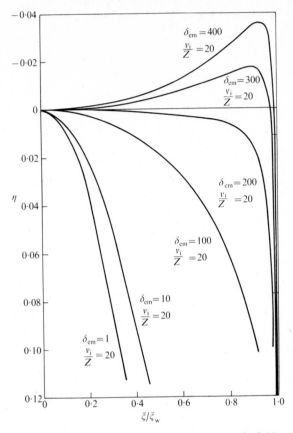

FIG. 3.2. The potential inversion which occurs at strong magnetic fields such that $\delta_{em} > (\nu_i/Z + \frac{3}{2})T_e/T_i)$ shown by varying δ_{em} at fixed $\nu_i/Z = 20$ and $T_e/T_i = 10$.

3.2. The positive column in a magnetic field at high pressures and magnetic field strengths

We have seen how the existence of an axial magnetic field can be regarded as an equivalent increase in pressure so far as the radial motion of particles is concerned. At high pressures and magnetic field intensities, it is a valid approximation to ignore inertial terms in the equations of motion. This implies that there will have to be a modified boundary condition at the wall, but otherwise we can proceed as before but treating both particles as magnetized.

As previously, we assume quasi-neutrality $n_e = n_i = n$ and the particle radial velocities must be equal so we set $v_{er} = v_{ir} = v$. Also we assume azimuthal homogeneity, that is, $\partial/\partial\theta$ of all quantities vanishes. Then the

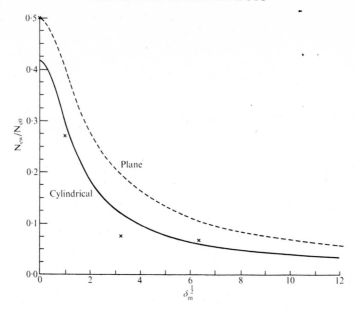

FIG. 3.3. The ratio of the wall electron density n_{ew} to the central electron density n_{e0} as a function of the magnetic field parameter $\delta_m = \delta_{em} + \delta_i$ for plane and cylindrical geometry compared with measurements.

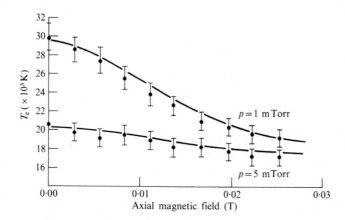

FIG. 3.4. The influence of an axial magnetic field on the electron temperature measured in mercury for two different pressures compared with theory. The theory is based on eqn (3.8) and the experiments use a microwave radiometer technique, $r_w = 13.5$ mm.

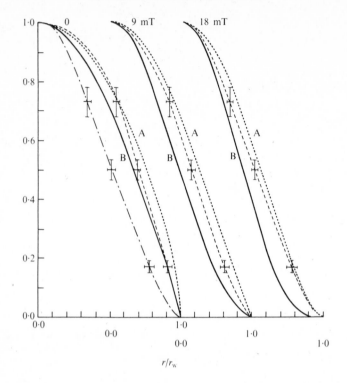

r/r_w

FIG. 3.5. The influence of an axial magnetic field on the radial electron distribution as seen in the visible light output. The theoretical solid curves are A for electron generation proportional to electron density n_e and B for generation proportional to n_e^2. The measured curves are dashed and the curve measured at $0.018 \, T$ is shown also at $B = 0$ to show the magnitude of the effect.

equations of motion are

$$\nu_i \mathbf{v}_i = -\frac{e}{M}\frac{d\phi}{dr}\hat{\mathbf{r}} + \frac{e}{M}\mathbf{v}_i \times \mathbf{B} - \frac{k_B T_i}{Mn}\frac{dn}{dr}\hat{\mathbf{r}}, \qquad (3.9)$$

$$\nu_e \mathbf{v}_e = \frac{e}{m}\frac{d\phi}{dr}\hat{\mathbf{r}} - \frac{e}{m}\mathbf{v}_e \times \mathbf{B} - \frac{k_B T_e}{mn}\frac{dn}{dr}\hat{\mathbf{r}}, \qquad (3.10)$$

which separate to give

$$v_{i\theta} = -\frac{\omega_{ci}}{\nu_i}v, \qquad v_{e\theta} = \frac{\omega_{ce}}{\nu_e}v, \qquad (3.11)$$

$$v\left(\nu_i + \frac{\omega_{ci}^2}{\nu_i}\right) = -\frac{e}{M}\frac{d\phi}{dr} - \frac{k_B T_i}{Mn}\frac{dn}{dr},$$

$$v\left(\nu_e + \frac{\omega_{ce}^2}{\nu_e}\right) = \frac{e}{m}\frac{d\phi}{dr} - \frac{k_B T_e}{mn}\frac{dn}{dr}.$$

Solving for $(1/n)(dn/dr)$ and $d\phi/dr$ gives

$$\frac{1}{n}\frac{dn}{dr} = -\frac{eBv}{k_B(T_e + T_i)}\left(\frac{\nu_i}{\omega_{ci}} + \frac{\omega_{ci}}{\nu_i} + \frac{\nu_e}{\omega_{ce}} + \frac{\omega_{ce}}{\nu_e}\right), \tag{3.12}$$

$$\frac{d\phi}{dr} = \frac{vB}{T_e + T_i}\left\{-T_e\left(\frac{\omega_{ci}}{\nu_i} + \frac{\nu_i}{\omega_{ci}}\right) + T_i\left(\frac{\omega_{ce}}{\nu_e} + \frac{\nu_e}{\omega_{ce}}\right)\right\}. \tag{3.13}$$

The potential inversion that we noted for intermediate fields when the electrons were magnetized but not the ions, and which occurs for fields such that

$$\omega_{ce}\omega_{ci} > \frac{T_e}{T_i}\nu_e\nu_i,$$

does not persist at even higher fields if $T_e\mu_i > T_i\mu_e$.

Substituting (3.12) into the continuity equation

$$\frac{1}{r}\frac{d}{dr}(rnv) = Zn$$

gives

$$\frac{1}{r}\frac{d}{dr}\left(r\frac{dn}{dr}\right) + \frac{neZB}{k_B(T_e + T_i)}\left(\frac{\nu_i}{\omega_{ci}} + \frac{\omega_{ci}}{\nu_i} + \frac{\nu_e}{\omega_{ce}} + \frac{\omega_{ce}}{\nu_e}\right) = 0, \tag{3.14}$$

with the Bessel-function dependence of the Schottky theory modified by the ambipolar diffusion coefficient D_{a0} (previously written D_a, eqn (2.20)) becoming D_{am}.

Figs 3.6(a) and (b) show measurements of the ambipolar diffusion

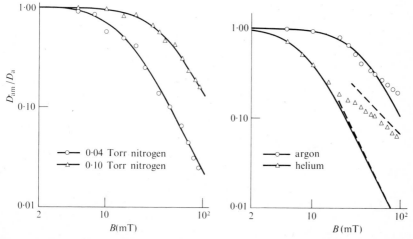

FIG. 3.6. The ambipolar diffusion coefficient in a magnetic field as a function of magnetic field in nitrogen at pressures 0·04 Torr and 0·10 Torr and in argon and helium at 0·04 Torr. Lines corresponding to $D_{am} \propto 1/B^2$ and $1/B$ are shown and the theoretical curve corresponds to eqn 3.14.

coefficient made in afterglows in nitrogen, helium, and argon by following the decay in mean electron density as described in § 5.10. The variation of D_{am} with B is shown, and these results (Ventrice and Brown 1972) indicate agreement with (3.14) throughout the range of B, i.e. up to 0·1 T in nitrogen. The other two gases show a departure from the $1/B^2$ dependence predicted by (3.14) at high fields and indicate rather a $1/B$ dependence. This effect which has been extensively studied has its origin in the excitation of waves in inhomogeneous plasmas in a magnetic field and will be discussed in Chapter 10.

3.3. Diamagnetism of a plasma column

One further bulk property of a plasma column which has been of interest over the years (Bohr 1911; van Leewen 1919; Steenbeck 1935), and which can be measured with some accuracy, is the diamagnetic moment which is defined by

$$m_d = -\pi e \int_0^{r_w} r^2 n_e(r) v_\theta(r)\, dr. \tag{3.15}$$

Using $v_\theta(r) = (\omega_{ce}/v_e)v_r(r)$ corresponding to electrons alone being magnetized, m_d can be expressed as an integral function of quantities which can be derived theoretically. Approximate forms can be found at low and high magnetic fields. At low pressures and low magnetic fields $\int nv_r\, dr$ will be a numerical multiple of the central number density n_0, the boundary radial velocity, and the discharge radius. Thus

$$m_d = -P\pi e r_w^3 n_0 \left(\frac{k_B T_e}{M}\right) \frac{\omega_{ce}}{v_e},$$

where P is a number less than unity.

Written in terms of the number per unit length $N_l = 2\pi \int nr\, dr$ this becomes

$$m_d = -QN_l r_w p \left(\frac{k_B T_e}{M}\right)^{\frac{1}{2}} \left(\frac{e^2 B}{mv_e/p}\right) \quad \text{(where } Q \text{ is a constant)}$$

$$= -\left\{ SN_l k_B T_e B \bigg/ \frac{Z}{p}(T_e) \right\} \frac{v_e}{p},$$

where S is a constant to be found from the plasma balance equation (3.8). Thus m_d increases at first approximation linearly with B.

At high fields the ion motion must be included, and using eqns (3.11) and (3.12) and adding the electron and ion azimuthal velocities to find the azimuthal current density,

$$nev_\theta = (v_{i\theta} + v_{e\theta})ne,$$

gives

$$m_d = -\pi \int nevr^2 \left(\frac{\omega_{ci}}{\nu_i} + \frac{\omega_{ce}}{\nu_e} \right) dr,$$

which, using eqn (3.12), gives

$$m_d = -\frac{k_B(T_e + T_i)}{B} \left(\frac{\omega_{ci}/\nu_i + \omega_{ce}/\nu_e}{\omega_{ci}/\nu_i + \nu_i/\omega_{ci} + \omega_{ce}/\nu_e + \nu_e/\omega_{ce}} \right) \int \pi r^2 \, dr \qquad (3.16)$$

$$\sim -\frac{k_B(T_e + T_i)}{B} N\pi r_w^2 \text{ for large } B, \qquad (3.17)$$

a result given by Allis and Gordon (1957) and Alfvén and Falthammar (1963).

It should be noted that a uniform bounded electron gas would be expected to have zero magnetic susceptibility because of the circulating wall current which results from reflection of those electrons whose orbits intersect the walls. However, the wall density rapidly drops to zero as B increases and the wall current 'paramagnetism' becomes negligible.

Detailed experimental measurements of the diamagnetic moment have been made by Forrest and Franklin (1968) by modulating a mercury positive column. They also measured independently electron temperature and number density. Their results are summarized in Fig. 3.7 showing that the full results of eqn (3.16) are needed to adequately describe the diamagnetism at low pressures.

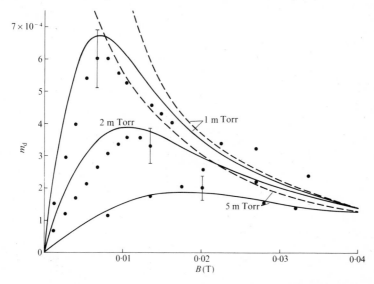

FIG. 3.7. The magnitude of the diamagnetic moment m_d of a cylindrical plasma as a function of the magnetic field showing a comparison between experiment in mercury at three pressures and theory. Eqn (3.16), full line; eqn (3.17), dashed line.

3.4. The Hall effect in the positive column

The effect of a transverse magnetic field on a positive column depends on its magnitude. If it is sufficiently large all particles are constrained by it to make collisions with the wall, and the effect is gross with virtual cathodes forming on the wall (see Francis 1957, p. 169); however, for weak fields it is possible to treat the problem by the methods of the last two chapters and also to make experimental measurements.

For simplicity consider a plane model with the magnetic field in the y-direction and the column axis in the z-direction. Only the electrons are assumed to be influenced by the magnetic field and their equation of motion becomes, for $\nu_e \gg Z$,

$$m\mathbf{v}_e \nu_e = -e(\mathbf{E} + \mathbf{v}_e \times \mathbf{B}) - \frac{1}{n}\nabla(nk_B T_e),$$

which separates into

$$m\nu_e v_{ez} = -eE_z$$

and

$$m\nu_e v_{ex} = -e(E_x - v_{ez}B_y) - \frac{k_B T_e}{n}\frac{dn}{dx}.$$

Ignoring electron inertia and combining with the ion equation of motion ($T_i = 0$),

$$M\left(v_{ix}\frac{dv_{ix}}{dx} + Zv_{ix} + \nu_i v_{ix}\right) = eE_x$$

and the continuity equation,

$$\frac{d}{dx}(nv_{ix}) = \frac{d}{dx}(nv_{ex}) = Zn,$$

gives normalized equations

$$\frac{d\xi}{du} = \frac{1-u^2}{1+u^2\,\delta + u\varepsilon_H}, \qquad (3.18)$$

$$\frac{d\eta}{du} = \frac{u(1+\delta) + \varepsilon_H u^2}{1+u^2\,\delta + \varepsilon_H u}, \qquad (3.19)$$

where the normalizations are as previously and in addition

$$\varepsilon_H = -\frac{ev_{ez}B}{Z(k_B T_e M)^{\frac{1}{2}}}.$$

The distributions cease to be symmetric about the midplane; in physical terms this is because of the $\mathbf{j} \times \mathbf{B}$ force due to the discharge current.

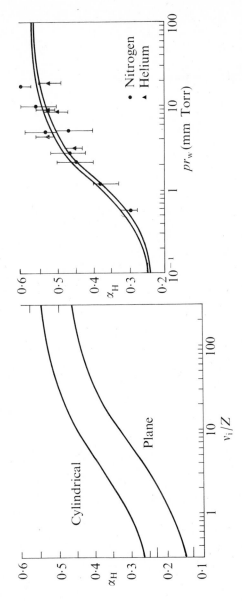

Fig. 3.8. The Hall coefficient $R_H \equiv \alpha_H/\bar{n}e$ of a positive column as a function of the collision parameter ν_i/Z in cylindrical and plane geometry. Measurements of α_H as a function of pressure in helium and nitrogen compared with theory.

Now the Hall coefficient in this situation can conveniently be defined in terms of the difference ΔV_H in potential at opposite points on the walls by

$$\Delta V_H = \alpha_H . 2L . v_{ez} . B. \tag{3.20}$$

The Hall coefficient R_H is then

$$R_H = \frac{\Delta V_H}{2L} . \frac{2L}{\displaystyle\int nev_{ez}\,dx . B} = \frac{\alpha_H}{\bar{n}e}.$$

Treating ε_H as a small parameter <1, ξ and η can be expanded as power series in ε_H,

$$\xi = \xi_0(u) + \varepsilon_H \xi_1(u) + \varepsilon_H^2 \cdots ,$$
$$\eta = \eta_0(u) + \varepsilon_H \eta_1(u) + \varepsilon_H^2 \cdots ,$$

and then to lowest order in ε_H, α_H is $\eta_1(1)/\xi_0(1)$. Similar expressions result in cylindrical geometry. The free-fall model of the positive column can also be analysed in the same spirit, and results have been given by Ecker and Kanne (1964).

Fig. 3.8 shows α_H as a function of v_i/Z; a comparison is also made of the values measured by Anderson (1964) with theory for nitrogen and helium over a range of pressure multiplied by radius.

4

EFFECTS OF SPACE CHARGE

4.1. The influence of a physical boundary to a plasma

SO FAR in treating the positive column we have applied, at the outer boundary or wall, the condition of single-valuedness which is essentially mathematical in origin but would be expected on physical grounds.

If conditions at the wall are to be correctly included then we need to take into account the different properties of ions and electrons. Under most circumstances the wall consists of an insulating material, and recombination of the charged particles takes place there on a time scale which will not enter into steady-state considerations. This implies that there must be equality of the fluxes of both types of particle to the wall so that

$$\Gamma_e = \Gamma_i. \tag{4.1}$$

If the electron and ion masses, temperatures, and collision frequencies were the same it would be possible to satisfy this condition with $n_e = n_i$ up to the wall; however, there is, in general, a large difference in mass, and the electron temperature is set by the required ionization rate while the ion temperature is close to the gas temperature and determined by thermal considerations. Near the wall the electron flux is largely determined by the electron random motion, and classical kinetic theory would suggest $n_{ew}\bar{c}_e/4$, where n_{ew} is the electron density at the wall and $\bar{c}_e$ is the mean random speed. The ion flux, on the other hand, is apparently determined by directed motion at low pressures, since we have been using a condition $u_w = 1$, or $v_i = (k_B T_e/M)^{\frac{1}{2}}$, which is greater than the random flux by a factor $(\pi T_e/8 T_i)^{\frac{1}{2}}$. This implies that there must be a region in which the number densities of electrons and ions are significantly different if an equal flux condition is to be achieved throughout the column. A difference in charge density implies the existence of a space-charge field which will itself influence the charged-particle motion in this region.

4.2. A simple model for the space-charge sheath at low pressure

One can attempt to set up a model for this region which is usually referred to as a space-charge sheath. The simplest model, some of the features of which we will have to justify *a posteriori*, is based on the following assumptions:

(1) the wall field causes the ions to reach the plasma–sheath boundary with a directed velocity v_{i0};

(2) v_{i0} is large compared with $\bar{c}_i$, the ion thermal speed;

(3) the plasma–sheath boundary is characterized by equality of particle densities and density gradients;

(4) there are no collisions, either ionizing or momentum-transferring, within the sheath.

Fig. 4.1 indicates the expected variations of number density and velocity within the sheath region. The appropriate equations are then

$$n_i v_i = n_0 v_{i0}, \qquad \text{ion continuity;} \qquad (4.2)$$

$$M v_i \frac{\mathrm{d}v_i}{\mathrm{d}x} = -\frac{e}{\mathrm{d}x}\frac{\mathrm{d}V}{\mathrm{d}x}, \qquad \text{ion momentum;} \qquad (4.3)$$

$$\frac{k_B T_e}{n_e}\frac{\mathrm{d}n_e}{\mathrm{d}x} = eV, \qquad \text{electron motion;} \qquad (4.4)$$

$$\frac{\mathrm{d}^2 V}{\mathrm{d}x^2} = \frac{e}{\varepsilon_0}(-n_i + n_e), \qquad \text{Poisson's equation;} \qquad (4.5)$$

with boundary conditions $n_i = n_e = n_0$ at $x = 0$, and also $V = 0$ and $\mathrm{d}V/\mathrm{d}x = \mathrm{d}n_i/\mathrm{d}x = \mathrm{d}n_e/\mathrm{d}x = 0$ at $x = 0$, the origin being taken at the plasma–sheath boundary.

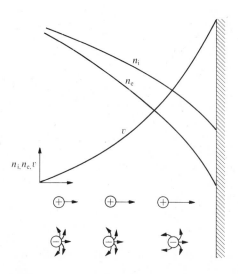

FIG. 4.1. Schematic diagram showing the expected variation of electron and ion densities n_e and n_i and the ion drift velocity v near an insulating wall. Both ion and electron velocity distributions are also indicated schematically.

Using normalizations as previously, we set

$$\eta = -\frac{eV}{k_B T_e}, \qquad N_e = \frac{n_e}{n_0}, \qquad N_i = \frac{n_i}{n_0}, \qquad u_i = \frac{v_i}{c_s}.$$

Eqns (4.2), (4.3), and (4.4) become

$$N_i u_i = u_{i0}, \qquad u_i \frac{du_i}{dx} = \frac{d\eta}{dx}, \qquad N_e = e^{-\eta},$$

whence

$$\frac{d^2\eta}{dx^2} = \frac{n_0 e^2}{\varepsilon_0 k_B T_e}(N_i - N_e) = \frac{1}{\lambda_D^2}\left(\frac{u_{i0}}{u_i} - e^{-\eta}\right)$$

$$= \frac{1}{\lambda_D^2}\left\{\left(1 + \frac{2\eta}{u_{i0}^2}\right)^{-\frac{1}{2}} - e^{-\eta}\right\}. \qquad (4.6)$$

The right-hand side can be expanded as a power series in η if η is less than both 1 and $u_{i0}^2/2$,

$$\frac{1}{\lambda_D^2}\left\{\eta\left(1 - \frac{1}{u_{i0}^2}\right) + \frac{\eta^2}{2}\left(\frac{3}{u_{i0}^4} - 1\right) + \ldots\right\}$$

This shows that η will be a monotonically increasing (i.e. non-oscillatory) function of distance if $u_{i0} > 1$, i.e. $v_{i0} \geqslant c_s$ and that the scale length for exponential increase of η is λ_D, the Debye length. The condition $v_{i0} \geqslant c_s$ is usually referred to as the Bohm criterion (1949) and interestingly, for equality, coincides with the boundary condition which we have been applying in the plasma approximation. Indeed the wall condition suggested earlier of equality of electron random and ion directed fluxes is, in normalized terms,

$$N_{ew} = \left(\frac{2\pi m}{M}\right)^{\frac{1}{2}},$$

which leads to values of the potential at the wall η_w ranging between 2·84 for hydrogen and 5·49 for mercury, so that in this model the space-charge region would be expected to extend for several Debye lengths.

Returning to eqn (4.6) we examine the behaviour of the solution more closely, and in particular proceed from the wall inwards rather than the plasma outwards. If we specify η_w and integrate once to obtain the field, it is possible then to proceed to find $\eta(x)$ numerically. This process yields a solution for η which tends asymptotically to zero but does not pass through it for finite x. Thus we cannot specify a joining of plasma and sheath at a point but rather should require that the sheath solution joins smoothly to the plasma for large $\xi = (x_w - x)/\lambda_D$. More precisely we require $\eta \to 0$, $d\eta/d\xi \to 0$ as $\xi \to \infty$. If now we examine the solutions of the form

$$\eta = \frac{A}{\xi^n} \qquad (4.7)$$

and consider (4.6) for small η we have

$$\frac{n(n+1)A}{\xi^{2+n}} = 1 - \frac{A}{\xi^n}\frac{c_s^2}{v_{i0}^2} + \frac{3}{2}\frac{A^2}{\xi^{2n}}\left(\frac{c_s^2}{v_{i0}^2}\right) \cdots - \left\{1 - \frac{A}{\xi^n} + \frac{A^2}{2\xi^{2n}} \cdots\right\}.$$

Clearly the terms in ξ^{-n} on the right-hand side must cancel giving

$$v_{i0}^2 = c_s^2 \qquad (4.8)$$

and thus the terms in ξ^{-2n} must equal that in $\xi^{-(n+2)}$. This requires $n = 2$, and this in turn gives $A = 6$. The value η_w is given by

$$\frac{n_0 \bar{c}_e}{4} \exp(-\eta_w) = n_0 v_0$$

or

$$\eta_w = \ln\frac{\bar{c}_e}{4v_0} = \tfrac{1}{2}\ln\frac{M}{2\pi m} \qquad (4.9)$$

as before. Since eqn (4.6) involves ξ only in the differential coefficient, η is undefined to an arbitrary constant and thus the form of the potential is given near the plasma–sheath boundary rather than its absolute value. Alternatively, we can integrate (4.6) from a large value of ξ requiring $\eta \sim 6/\xi^2$ and find the universal sheath-potential–distance curve, which is terminated where the sheath potential reaches the value required by the particular ion species $\tfrac{1}{2}\ln(M/2\pi m)$.

It can be seen that the sheath thickness (i.e. the distance from the wall to the plasma) is ill defined. However, a lower bound can be put to it by considering the form of the solution near the wall, namely,

$$\frac{d^2\eta}{d\xi^2} \sim \frac{1}{(2\eta)^{\frac{1}{2}}},$$

which integrates to give

$$\tfrac{4}{3}(\eta_w)^{\frac{3}{4}} \sim 2^{\frac{3}{4}}(\xi_p - \xi_w)$$

or

$$(\xi_p - \xi_w) \sim \frac{4}{3}\left(\frac{\eta_w}{2}\right)^{\frac{3}{4}} = 1 \cdot 74 \text{ for hydrogen, and } 2 \cdot 84 \text{ for mercury.} \qquad (4.10)$$

These values correspond to a space-charge-limited region containing ions alone, and therefore the Child (1911)–Langmuir (1913) relation (4.10) applies.

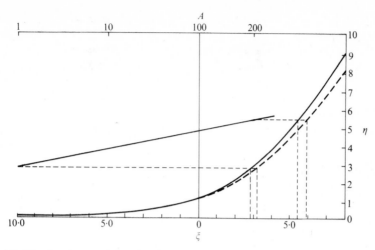

FIG. 4.2. The sheath potential $\eta = eV/k_B T_e$ as a function of normalized distance for plane (dashed) and cylindrical (full) geometries. For an ion of atomic number A a horizontal line drawn through the point where the abscissa through A meets the inclined line cuts the potential curves at the wall position.

Typical sheath thicknesses are quoted as being somewhat larger, e.g. 10–20 Debye lengths. Fig. 4.2 gives a curve for η versus ξ and also a means of determining η_w as a function of the ion mass.

The Bohm criterion stated as the directed velocity of the ions having to exceed the speed for ion acoustic waves for there to be a transition from plasma to space-charge sheath is formally analogous to the condition in plane flow of a fluid for a smooth transition in a convergent–divergent nozzle (see Persson 1962). This 'irreversible' transition from plasma to sheath is further explored in §7.9, where the propagation of ion waves. through such a region is discussed.

In physical terms it can be seen that the electrostatic forces which arise when there is an imbalance of densities of space charge are so powerful that they can be overcome only if there is a sufficiently large directed energy (for particles of one sign) and that even in this case the scale length for significant departures from charge neutrality is the Debye length.

4.3. Generalizing the Bohm criterion

The model equations for the sheath given above assumed that the ions arrived at the edge of the plasma with some directed speed v_{i0}, and the requirement that the electric field grew monotonically in the sheath gave the Bohm criterion $Mv_{i0}^2 = k_B T_e$. It is implicit in this model of the positive column that the ions have a common radial velocity and so the ion speed can vary continuously through plasma and sheath. On the other hand, in the free-fall model with ions entering the sheath with a speed determined by the potential difference each has fallen through from its point of generation there will be a distribution of ion energies at any point in the plasma. Since this is physically more realistic at low pressures it is of interest to seek to generalize the Bohm criterion to take account of a distribution of ion velocities.

Let us suppose that in the plasma at the sheath edge the ion distribution is $f(v)$. As an ion with radial velocity v falls through the sheath it acquires a velocity v', where $v'^2 = v^2 + (2eV/M)$, and V is the potential at a point in the sheath relative to the plasma. The ion distribution function becomes $g(v')$, where g is given by conservation of particles $v'g(v')\,dv' = vf(v)\,dv$, assuming a plane boundary, and the particle density n' is $\int g(v')\,dv'$, which can be written

$$\int_0^\infty \frac{v}{v'} f(v)\,dv$$

or

$$\int_0^\infty \frac{vf(v)\,dv}{\{v^2 + (2eV/M)\}^{\frac{1}{2}}}.$$

The sheath equation then becomes

$$-\frac{d^2 V}{dx^2} = \frac{en_0}{\varepsilon_0} \left[\int_0^\infty \frac{vf(v)\,dv}{\{v^2 + (2eV/M)\}^{\frac{1}{2}}} - \exp\left(-\frac{eV}{k_B T_e}\right) \right],$$

and the right-hand side may be written

$$\int_0^\infty f(v)\,dv - \frac{eV}{M}\int_0^\infty \frac{1}{v^2} f(v)\,dv + O\left(\frac{eV}{Mv^2}\right)^2 - 1 + \frac{eV}{k_B T_e} - O\left(\frac{eV}{k_B T_e}\right)^2,$$

expanding in powers of V provided that the function f satisfies appropriate conditions. Since f is a distribution function $\int_0^\infty f(v)\,dv$ exists; but the requirement that $\int_0^\infty (1/v^2)f(v)\,dv$ exist means that either $f(v) = 0$ for $v < v_c$, where v_c is some positive cut-off velocity, or that

$$f(v) \sim v^\alpha \quad \text{for small } \alpha, \quad \text{where} \quad \alpha > 1. \qquad (4.11)$$

Under these circumstances then the criterion for the existence of a space-charge sheath would become

$$\frac{M}{k_B T_e} = \int_0^\infty \frac{1}{v^2} f(v)\,dv, \qquad (4.12)$$

a result given by Harrison and Thompson (1959). The drift velocity $v_D = \int_0^\infty vf(v)\,dv$ so that the condition becomes $Mv_D^2 = C_f k_B T_e$, where

$$C_f = \int_0^\infty \frac{1}{v^2} f(v)\,dv \left(\int_0^\infty vf(v)\,dv\right)^2$$

and is a number of order unity.

As an example let us consider the situation for the free-fall model in plane geometry with generation proportional to electron number density. This is a case treated by Tonks and Langmuir (1929) and whose solution has been given in a convenient form by Harrison and Thompson (1959). In terms of the normalized variables, $\eta = -(eV/k_B T_e)$, $\xi = xZ/c_s$, it is straightforward to show that $f(v)$ at the plasma boundary is given by $f(v) \propto n(\xi)(d\xi/d\eta)$, where v is related to ξ the point of generation by

$$v = \sqrt{2}c_s\{\eta(\xi_0) - \eta(\xi)\}^{\frac{1}{2}},$$

ξ_0 being the singular point where $d\eta/d\xi \to \infty$. Thus

$$f(v) \propto \eta^{-\frac{1}{2}} - 2e^{-\eta}D(\eta^{\frac{1}{2}})$$

where D is the Dawson function defined by $D(x) = \int_0^x \exp t^2\, dt$. Now $f(v) \sim (1/2\eta_0^{\frac{1}{2}})(\eta_0 - \eta)$ for $\eta \to \eta_0$ so that $f(v) \sim v^2$ and the requirement (4.11) is satisfied. Fig. 4.3 shows $f(v)$ on linear and logarithmic scales.

It is possible to show that for $f(v) \propto v^{2n}$, $0 < v < v_m$, $f(v) = 0$, $v > v_m$ and $n \geqslant 1$ the criterion becomes

$$Mv_m^2 \frac{(2n-1)}{(2n+1)} = k_B T_e.$$

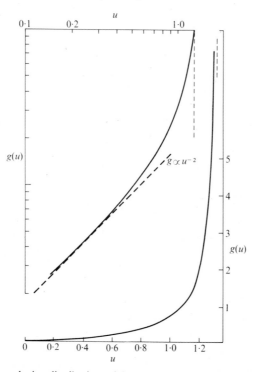

FIG. 4.3. The ion velocity distribution $g(u)$ as a function of the normalized velocity u leaving the plasma and entering the sheath for a plane free-fall discharge. The upper curve on a logarithmic scale shows $g(u) \propto u^{-2}$ for $u \to 0$. The lower curve is on a linear scale.

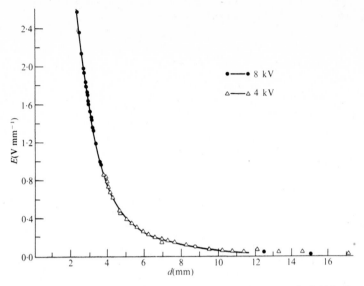

FIG. 4.4. Electron-beam measurements at 4 kV and 8 kV of the electric field in the sheath adjacent to a plane insulating surface in a low-pressure discharge compared with free-fall theory.

In the limit for large n a δ-function velocity distribution results and (4.8) is recovered.

A comparison between the free-fall model and experiment is provided by the work of Goldan (1970) at low pressure (0·85 mTorr) in an argon discharge using a transverse electron beam to measure the electric field in the sheath. Typical results are shown in Fig. 4.4.

4.4. Joining active plasma and sheath

There is some inconsistency in the model derived above as it is based on a uniform-density zero-field plasma in the region $x < 0$ which is scarcely consistent with the ions having a directed velocity. Also in our model of the positive column we have an electric field and a number-density gradient which at low pressures becomes infinite at the boundary. This disjoint nature of conditions led to the speculation that there might be a presheath region in which the electric number densities, fields, and radial velocities adjusted themselves to render the situation self-consistent. There is some interest in discovering the extent of this region or alternatively, the distance of field penetration in the plasma.

Some progress can be made towards answering these questions by considering the form of the equations which arise when the plasma approximation $n_e = n_i$ is not made. Then one has equations which are a simple modification of (2.9), (2.10), and (2.11) (p. 27) with

$N_e = n_e(\xi)/n_e(0)$ and $N_i = n_i(\xi)/n_e(0)$, namely,

$$\frac{d}{d\xi}(N_i u) = N_e, \tag{2.9a}$$

$$u\frac{du}{d\xi} + u\left(\frac{N_e}{N_i} + \frac{\nu_i}{Z}\right) = \frac{d\eta}{d\xi}, \tag{2.10a}$$

$$N_i u\, \delta_e = -\frac{dN_e}{d\xi} - N_e\frac{d\eta}{d\xi}, \tag{2.11a}$$

and there is now in addition Poisson's equation,

$$\alpha^2 \frac{d^2\eta}{d\xi^2} = N_i - N_e, \tag{4.13}$$

where

$$\alpha^2 = \frac{\varepsilon_0 Z^2 M_i}{n_e(0)e^2} \equiv \frac{Z^2}{\omega_{pi}^2}\frac{n_i(0)}{n_e(0)}.$$

Subsequently $n_e(0)$ and $n_i(0)$ are written n_{e0}, n_{i0}.

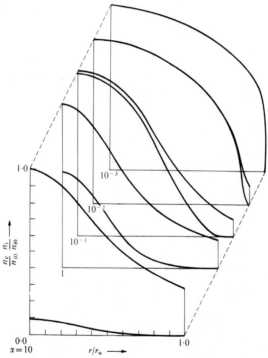

Fig. 4.5. Radial distributions of electron and ion density normalized to the central *ion* density n_{i0} for different values of parameter $\alpha = (\varepsilon_0 Z^2 M_i/n_{e0}e^2)^{\frac{1}{2}}$ with $\nu_i = 0$, i.e. $\delta = 1$ showing growth of the sheath as α increases.

The equations cannot be solved analytically for $\alpha \neq 0$ so that recourse has to be had to computation, and since the plane problem is of little practical consequence it is more meaningful to examine the situation in cylindrical geometry. In order to begin the calculation working out from the centre as origin with boundary conditions

$$N_e = 1, \qquad u = 0, \qquad \eta = 0, \qquad \frac{d\eta}{d\xi} = \frac{dN_i}{d\xi} = \frac{dN_e}{d\xi} = 0,$$

it is necessary to specify $N_{i0} = n_{i0}/n_{e0}$. This can be found from the equations by expanding each variable in a Taylor's series at the origin and requiring the differential equations to be satisfied to give an equation for

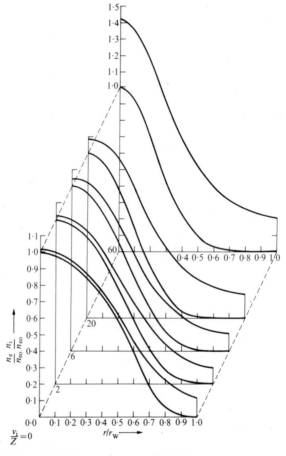

FIG. 4.6. Radial distributions of relative electron and ion densities for $\alpha = 0 \cdot 1$ as the ion collision parameter $\nu_i/Z = \delta - 1$ is varied, showing modest growth of the sheath as ν_i/Z increases.

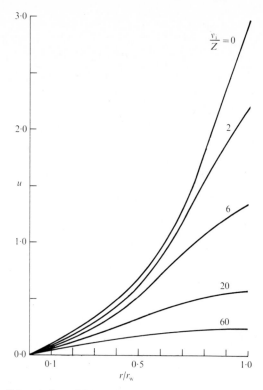

FIG. 4.7. The radial variation of ion radial velocity as v_i/Z is varied, showing that for $v_i/Z \geqslant 10$ the Bohm speed is not reached.

$f_0 \equiv N_{i0}$,

$$f_0^3 - f_0^2 - \alpha^2 \left(f_0 \frac{v_i}{Z} + \frac{3}{2} \right) = 0, \qquad (4.14)$$

which for small α ($f_0 \sim 1$) can be solved to give $f_0 \simeq 1 + \alpha^2(\frac{1}{2} + \delta_i)$, showing that there is an excess of positive charge of order α^2 at the centre. The integration can be carried out with either ξ or u as the independent variable. The flux balance condition at the wall is

$$N_i u = N_e \left(\frac{M}{2\pi m} \right)^{\frac{1}{2}} \quad \text{for} \quad \Gamma_e = \frac{n_e \bar{c}_e}{4},$$

and so the wall position can be regarded as a variable to be calculated, with the wall position ξ_w now being a function of α, δ_i, and δ_e. Care must be taken with the step length when integrating numerically in the region of the sheath and typical results are shown in Figs 4.5, 4.6, and 4.7, and tabulated in Tables 4.1 and 4.2.

TABLE 4.1 $\xi_w(\delta_i, \alpha)$

α \ δ_i	0	1	3	10	30	100	300	1000	
0	1·1094	0·8615	0·6602	0·4355	0·2760	0·1597	0·0948	0·0528	
10^{-3}	1·1327	0·8845	0·6833	0·4586	0·2985	0·1799	0·1121	0·0677	$\Gamma_e = \dfrac{n_e \bar{c}_e}{4}$
10^{-2}	1·2664	1·0153	0·8107	0·5764	0·4007	0·2676	0·1923		
10^{-1}	2·1896	1·8688	1·5685	1·1958	0·9211				
0	1·1094	0·8615	0·6602	0·4355	0·2760	0·1597	0·0948	0·0528	
10^{-3}	1·1333	0·8852	0·6841	0·4597	0·2998	0·1813	0·1133	0·0687	$\Gamma_e = \dfrac{n_e \bar{c}_e}{2}$
10^{-2}	1·2726	1·0226	0·8191	0·5858	0·4099	0·2756	0·1991		
10^{-1}	2·2598	1·9444	1·6413	1·2570	0·9710				

TABLE 4.2 $\eta_w(\delta_i, \alpha)$

α \ δ_i	0	1	3	10	30	100	300	1000	
0	6·36	6·71	7·07	7·60	8·13	8·75	9·32	9·93	
10^{-3}	6·36	6·72	7·09	7·64	8·21	8·86	9·47	10.15	$\Gamma_e = \dfrac{n_e \bar{c}_e}{4}$
10^{-2}	6·41	6·80	7·20	7·80	8·42	9·13	9·77		
10^{-1}	6·58	7·04	7·46	8·02	8·51				
0	6·36	6·71	7·07	7·60	8·13	8·75	9·32	9·93	
10^{-3}	7·06	7·42	7·78	8·33	8·90	9·56	10·17	10·86	$\Gamma_e = \dfrac{n_e \bar{c}_e}{2}$
10^{-2}	7·10	7·50	7·91	8·51	9·14	9·85	10·50		
10^{-1}	7·80	7·97	8·20	8·76	9·26				

Fig. 4.5 shows the influence of α, which can be related to the ratio discharge radius/Debye length, indicating that for $\alpha \gtrsim 10^{-2}$ the scale of the sheath is significant and that for $\alpha \gtrsim 1$ there are no regions of quasi-neutrality.

Fig. 4.6 shows for a fixed value of $\alpha = 0\cdot1$ the influence of the collision parameter δ_i and leads to the conclusion that the effect of collisions is to make the discharge less neutral and to concentrate the electrons in the centre of the discharge. This is less marked at smaller values of α. Fig. 4.7 gives the normalized ion radial velocity as a function of radius for $\alpha = 0\cdot1$ for different values of δ and shows that for δ significantly greater than 1 the Bohm criterion loses its meaning and the motion is essentially determined by collisions everywhere.

Tables 4.1 and 4.2 give values of the wall position $\xi_w(\delta_i, \alpha)$ and the wall potential for different values α and δ_i with $\delta_e = 0$ and for two values of the electron flux $n_e \bar{c}_e/4$ and $n_e \bar{c}_e/2$. The significance of these as limiting values is discussed in § 4.7. The gas was taken to be mercury vapour.

The process can be carried further to examine the consequences of non-zero α on the discharge parameters through the plasma balance

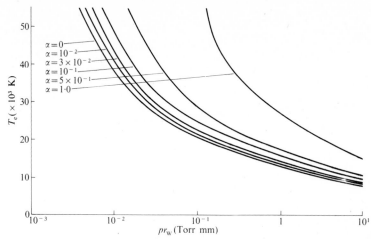

FIG. 4.8. The influence of space charge on the plasma balance equation relating the electron temperature T_e and pressure multiplied by radius, showing how for constant pr_w, T_e increases with α. The data are for mercury vapour.

equation, and, for instance, Fig. 4.8 shows the way in which the electron temperature in mercury is expected to vary with α as a function of pr_w. This can be expressed in terms of discharge parameters by converting number densities to currents, using appropriate similarity variables; Fig. 4.9 shows how at low current and low pressure T_e can become significantly dependent on current. This regime is sometimes referred to as a subnormal positive column.

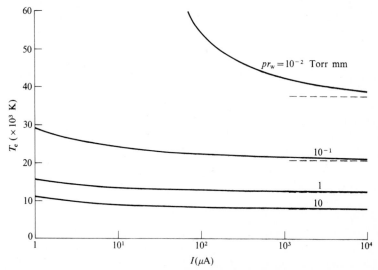

FIG. 4.9. The variation of electron temperature T_e as a function of the discharge current I in mercury with pr_w as a parameter. For a significant departure from the high current value of T_e the discharge is said to be subnormal.

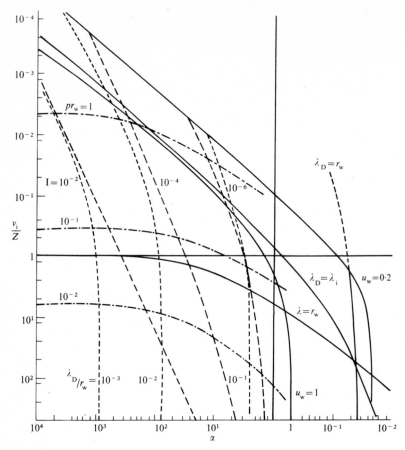

FIG. 4.10. The results of Figs 4.8 and 4.9 combined in a map for mercury vapour of curves of current I, ratio of Debye length to discharge radius λ_D/r_w, and pressure multiplied by radius pr_w in terms of the parameters v_i/Z and α. Limiting values are indicated by the solid lines.

Using these results it is possible, in principle, to construct a diagram relating the parameters α and δ to the similarity variables current I and pressure multiplied by radius pr_w. This has been done for mercury in Fig. 4.10, where also shown are limiting regime defined by (1) the ion mean free path λ_i being equal to the tube radius r_w, defining the boundary between collisionless and collisional discharges; (2) λ_i being equal to the Debye length λ_D, i.e. comparable with the 'sheath thickness', when the sheath becomes collisional; (3) λ_D being equal to r_w when the discharge ceases to be a plasma, and (4) $u_w = 0.2$, which corresponds to the ion radial velocity being so small that the ion temperature should be included in the equation of motion and particularly the boundary conditions.

4.5. Asymptotic analyses of the positive column

The results of the previous section are useful in determining the overall electrical characteristics of a discharge, but they do not give insight into the structure of the column and for this reason it is worthwhile, in a manner similar to that used in fluid boundary-layer theory, to carry out a matched asymptotic expansion procedure to join plasma to sheath. Such a treatment has been given by Franklin and Ockendon (1970) in plane geometry and cylindrical geometry for $\delta = 1$ and for the Tonks–Langmuir free-fall model and by Blank (1968) for the collision-dominated situation. It is interesting to contrast the two extreme cases. At low pressures in all cases it is necessary to introduce a transition layer to enable proper matching of sheath and plasma, and this layer is characterized by having a scale which varies as $\alpha^{\frac{4}{5}}$ in the fluid model and $\alpha^{\frac{8}{9}}$ in the free-fall model. Thus the presheath is of greater physical extent than the sheath and indeed on the scale of the sheath thickness extends throughout the discharge. Fig. 4.11 shows the variation of radial velocity computed directly and derived from the three approximate solutions.

On the other hand, it is possible to examine the situation when the collision frequency is so high that motion in the sheath is collision-dominated, while still ignoring generation within the sheath. Ion motion is then given by $Mv_i\nu_i = -e(dV/dx)$, $n_iv_i = n_0v_{i0} = J$, say, and the electron

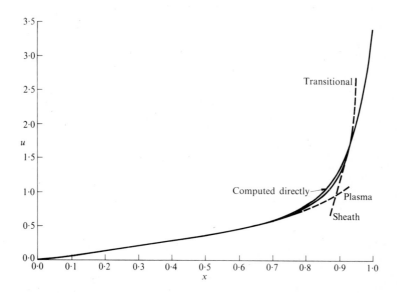

FIG. 4.11. The radial ion velocity computed for $\alpha = 0.0144$ compared with the asymptotic analytical plasma, sheath, and transitional solutions showing diagrammatically the need for the inclusion of a transitional layer.

density by $n_e = n_0 \exp(eV/k_B T_e)$. Poisson's equation becomes

$$\frac{d^2 V}{dx^2} = \frac{e}{\varepsilon_0 v_i} (-n_0 v_{i0} + n_e v_i),$$

using the ion mobility

$$\mu_i \frac{dV}{dx} \frac{d^2 V}{dx^2} = \frac{e}{\epsilon_0} \left\{ -J + n_0 \exp\left(\frac{eV}{k_B T_e}\right) \mu_i \frac{dV}{dx} \right\},$$

which integrates to

$$\mu_i \left(\frac{dV}{dx}\right)^2 = 2 \frac{e}{\varepsilon_0} \left\{ -Jx + n_0 \mu_i \frac{k_B T_e}{e} \left[\exp\left(\frac{eV}{k_B T_e}\right) - 1 \right] \right\}. \quad (4.15)$$

It can be readily shown that the appropriate power-series expansion is $V = -\beta x^{\frac{3}{2}}$, and matching terms it gives for the leading coefficient

$$\beta = \frac{2}{3}\left(\frac{eJ}{\mu_i \varepsilon_0}\right)^{\frac{1}{2}}.$$

It is clear then that the appropriate scaling of variables within the sheath is

$$\overline{n_{i,e}} = \frac{n_{i,e}}{n_0}\left(\frac{\lambda_D}{r_w}\right)^{\frac{2}{3}}, \qquad \bar{x} = \left(1 - \frac{x}{r_w}\right)\left(\frac{r_w}{\lambda_D}\right)^{\frac{2}{3}}, \qquad \bar{E} = \frac{eEr_w}{k_B T_e}\left(\frac{\lambda_D}{r_w}\right)^{\frac{2}{3}}$$

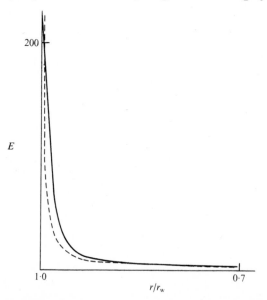

FIG. 4.12. The normalized radial electric field E near the wall in the presence of a collisional space-charge sheath. The 'plasma' solution (dashed) is joined smoothly by the sheath solution (full); $\lambda_D/r_w = 10^{-3}$ and $E \equiv eE_r r_w / k_B T_e$.

and that its thickness now is of the order of $\lambda_D^{\frac{2}{3}}$. A complete asymptotic treatment has been given by Blank, who has shown that the asymptotic form of the plasma balance equation is

$$\left(\frac{Z}{D_a}\right)^{\frac{1}{2}} r_w = 2.4048\left\{1+\frac{1}{\bar{J}}(E_w^2 - 2 \exp \bar{V}_w)\left(\frac{\lambda_D}{r_w}\right)^{\frac{2}{3}}\right\},$$

where $\bar{J}$ is the normalized ion flux. In this case the sheath and plasma solutions match without the need for an intermediate layer, as can be seen in Fig. 4.12.

It is natural to inquire whether the matched-asymptotic-expansion method can be applied to cover the range from the collisionless to the collisional sheath. In principle this is possible and it has been carried out for the closely related problem of the sheath around a probe in a plasma by Lam (1967). From his work it is possible to conclude that a multiple-layer structure is necessary in the transition region.

4.6. Finite ion temperature

One further step can be taken in relaxing assumptions made in setting up the model of the positive column and that is to allow the ion temperature to be non-zero. Considering plane geometry for simplicity the equations then become

$$\frac{d}{dx}(n_e v_e) = \frac{d}{dx}(n_i v_i) = Z n_e,$$

$$n_i v_i \nu_i + \frac{d}{dx}(n_i v_i^2) = n_i \frac{eE}{M} - \frac{k_B T_i}{M} \frac{dn_i}{dx},$$

$$n_e v_e \nu_e = -n_e \frac{eE}{m} - \frac{k_B T_e}{m} \frac{dn_e}{dx},$$

$$\frac{dE}{dx} = \frac{e}{\epsilon_0}(n_i - n_e).$$

Regarding these as a set of simultaneous first-order equations one can solve for each derivative in turn by Cramer's rule to yield equations like

$$\frac{dn_i}{dx} = \frac{2v_i Z n_e + n_i v_i \nu_i - n_i(eE/M)}{v_i^2 - (k_B T_i/M)}.$$

There is an apparent singularity where $v_i = (k_B T_i/M)^{\frac{1}{2}}$, as has been discussed by Friedman and Levi (1967), but now n_{i0}/n_{e0} can no longer be found by a Taylor expansion at the origin and the value must be such as to allow the solution to pass smoothly through the point where $v_i = (k_B T_i/M)^{\frac{1}{2}}$.

This can be achieved if simultaneously

$$\frac{eE}{M} = v_i \left(v_i + \frac{2n_e}{n_i} Z \right).$$

A means of dealing with this removable singularity numerically has been given by Crawford and Self (1967).

However, a more consistent treatment can be given, and one which involves no computational instability, if the ratio $\beta \equiv T_i/T_e$ is regarded as a small parameter and perturbation methods used. Confining ourselves to the collisionless case for the sake of illustration, i.e. putting $v_i = v_e = 0$, we seek to expand the variables as power series in β, e.g.

$$n_e = n_{e0} + \beta n_{e1} + \beta^2 n_{e2} + \cdots.$$

As we have already noted for the case $\beta = 0$ the singularity at the origin $x = 0$ can be dealt with by requiring all the variables to be non-singular there. Proceeding then in the same spirit and using the normalizations

$$f = \frac{n_i}{n_{e0}}, \qquad g = \frac{n_e}{n_{e0}}, \qquad h = \frac{v_i}{c_s}, \qquad y = \frac{Zx}{c_s}, \qquad \varepsilon = \frac{c_s eE}{Zk_B T_e},$$

the governing equations become

$$\frac{d}{dy}(hf) = g, \qquad \frac{dg}{dy} = -g\varepsilon,$$

$$\frac{d}{dy}(fh^2) + \beta \frac{df}{dy} = f\varepsilon, \quad \text{and} \quad \frac{d\varepsilon}{dy} = \frac{f - g}{\alpha^2},$$

where

$$\alpha^2 = \frac{\varepsilon_0 M Z^2}{n_{e0} e^2}.$$

The zero-order equations are

$$f_0' = \frac{2g_0 h_0 - f_0 \varepsilon_0}{h_0^2}, \qquad g_0' = -g_0 \varepsilon_0,$$

$$h_0' = \frac{f_0 \varepsilon_0 - g_0 h_0}{h_0 f_0}, \qquad \varepsilon_0' = \frac{f_0 - g_0}{\alpha^2},$$

with boundary conditions

$$f_0(0) = F_0, \qquad g_0(0) = 1, \qquad h_0(0) = 0, \qquad \varepsilon_0(0) = 0.$$

With these boundary conditions it can be readily be shown that f_0 and g_0 are even functions of y and h_0 and ε_0 odd.

The equations have solutions $f_0 \sim F_0 - \mu y^2$, $g_0 \sim 1 - (y^2/F_0^2)$ $h_0 \sim y/F_0$, $\varepsilon_0 \sim 2y/F_0^2$, where $F_0^2(F_0 - 1) = 2\alpha^2$ and μ has to be determined by examining higher-order terms and is found to be $(4F_0-3)/F_0(10F_0-9)$. The equations for f_1, g_1, h_1, and ε_1 are somewhat more complicated and have boundary conditions

$$f_1(0) = F_1, \qquad g_1(0) = 0, \qquad h_1(0) = 0, \qquad \varepsilon_1(0) = 0.$$

Ensuring non-singularity at the origin again gives F_1, which is found to be

$$F_1 = -\frac{F_0(F_0-1)(4F_0-3)}{(10F_0-9)(3F_0-2)},$$

while approximate forms for h_1 and ε_1 are

$$h_1 = -\frac{F_1}{F_0^2}y, \qquad \varepsilon_1 = \frac{F_1}{\alpha^2}y,$$

and thus for any given α^2 solutions for different β can be constructed by adding f_0 and βf_1 etc. Fig. 4.13 gives values of f_0, g_0, h_0, and ε_0 and f_1, g_1, h_1, and ε_1 for a range of values of α^2.

This difficulty emphasizes the fact already implied in much of the work in this chapter that a bounded plasma always 'knows' that is bounded and the effects of the boundedness make themselves felt at all points in the plasma on some scale. The influence of ion temperature on the plasma

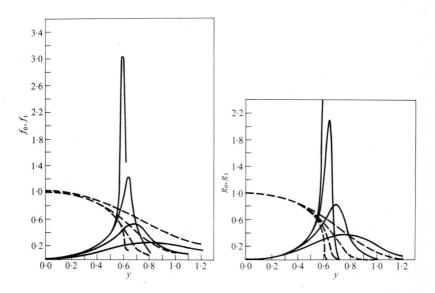

FIG. 4.13. Transverse distributions of normalized electron density g_0 (dashed), g_1 (full), and ion density f_0 (dashed), f_1 (full), for different values of the space-charge parameter α. The density correct to first order in $\beta \equiv T_i/T_e$, the ratio of ion temperature to electron temperature, is given by $g = g_0 + \beta g_1$ and $f = f_0 + \beta f_1$. The transverse normalized coordinate $y \equiv xZ/c_s$, and the values of α^2 from left to right are 10^{-5}, 10^{-4}, 10^{-3}, and 10^{-2}.

balance equation is most conveniently examined by a modification of the analysis given by Franklin and Ockendon (1970), giving for the plane case

$$y_w = [(\pi/2) - 1](1 + \beta)^{\frac{1}{2}}\{1 + 0 \cdot 438\alpha^{\frac{4}{5}}(1 + [\beta/2])^{\frac{3}{2}}(1 + \beta)^{-\frac{5}{4}}\}.$$

4.7. Detailed conditions adjacent to a wall

Since some treatments of plasma boundaries show the existence of a so-called diffusion layer (Su and Lam 1963), it is appropriate to include a discussion of this point. The need for this layer arises because the mathematical boundary condition is set that the ion number density at the wall should vanish, whereas the equality of fluxes condition implies a non-vanishing density. Whether one is more appropriate than another depends on the physical conditions at the surface and the mechanism by which recombination occurs. Ions arriving at the surface, if it were metallic, would recombine with electrons drawn locally from the metal, and this process would be rapid; on the other hand, if the surface were a perfect insulator, recombination would depend on electron mobility on the surface, and the ions would have a finite density corresponding to a finite lifetime. In any case the influence on the overall balance is small. Little if any experimental work appears to have been done to investigate the details of charged-particle behaviour on the walls, but there are some experiments on the lifetime of charges on insulating surfaces (Rosa 1973).

One point which we have so far not considered is the modification of the particle distribution functions by the existence of the wall. If the wall is perfectly absorbing, the fact that there is a net electron flux to the wall will modify the distribution function and therefore the simple value for the particle flux will need to be recalculated. Since typical wall potentials are several times $k_B T_e/e$, the modification within the plasma will be sufficiently small to allow us to neglect it, but as the boundary is approached the effect will be greater as a larger proportion of the electrons are not returned to the plasma by the potential barrier. This problem has been considered in connection with electrostatic probes by several authors and in considerable detail by Muller and Wahle (1970), who solved the Boltzmann equation directly. Their conclusions are that at high pressures there is little physical difference between vanishing density and a matching of random and directed flux conditions, while at lower pressures the error in taking the net electron flux to be $n_e \bar{c}_e/4$ is not large. This is summarized in Fig. 4.14, which shows

$$\alpha_e \equiv \left(\frac{4n_e v_e}{n_e \bar{c}_e}\right)_w$$

as a function of the ratio of ion mean free path to Debye length λ_i/λ_D when there is zero net current through the sheath, i.e. $n_i v_i = n_e v_e$. Muller

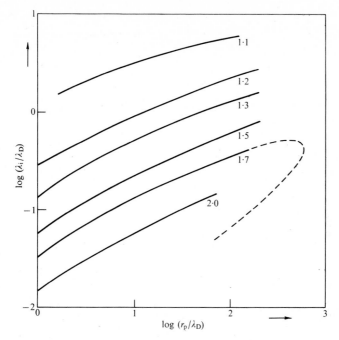

FIG. 4.14. The results of a kinetic-theory calculation of α_e the ratio of the electron wall flux to the uniform density random flux crossing a plane in one direction shown as functions of the ratio of ion mean free path λ_i to Debye length λ_D, and probe radius r_p to Debye length, for a spherical probe.

and Wahle also give α_e for situations in which there is a current flowing through the sheath which are of interest in probe theory but not needed in connection with the positive column.

A relatively straightforward argument can be produced to suggest that α_e should lie between 1 and 2. From kinetic theory the flux of electrons to the walls is

$$\Gamma_r = \frac{n_e \bar{c}_e}{4} - \frac{1}{2}\left(D_e \frac{dn_e}{dx} - n_e \mu_e E \right),$$

and that away from the wall

$$\Gamma_1 = \frac{n_e \bar{c}_e}{4} + \frac{1}{2}\left(D_e \frac{dn_e}{dx} - n_e \mu_e E \right),$$

where the first terms apply in a uniform plasma and the second and third give the net flux due to non-uniformity. If the wall has a 'reflection coefficient' r defined by $\Gamma_1 = r\Gamma_r$, then

$$\frac{n_e \bar{c}_e}{4}(1-r) = -\frac{(1+r)}{2}\left(D_e \frac{dn_e}{dx} - n_e \mu_e E \right)$$

or

$$\Gamma_r = \frac{n_e \bar{c}_e}{4}\left(1 + \frac{1-r}{1+r}\right) = \frac{n_e \bar{c}_e}{4}\left(\frac{2}{1+r}\right)$$

$$\Gamma_{net} = \Gamma_r - \Gamma_1 = 2\left(\frac{n_e \bar{c}_e}{4}\right)\frac{(1-r)}{(1+r)}. \qquad (4.16)$$

As r ranges between 0 and 1, corresponding to perfectly absorbing to perfectly reflecting, α_e then ranges from 2 to 1, and one would expect the effective value of r to be largest for large λ_i (i.e. near collisionless conditions) and to decrease as λ_D increases. These are the general features of Fig. 4.14.

4.8. Axial sheaths in a positive column

The arguments which were used to predict the existence of space-charge sheaths near the boundary walls of a positive column can with some modification be applied to the regions adjacent to the bounding electrodes. In particular, the situation near the cathode of a cold-cathode discharge is complicated by the need for the electrons emitted from the cathode first to acquire directed energy and then to be scattered so as to form a swarm. Details depend on the particular gas and material of the cathode. For a comprehensive treatment see von Engel (1965). On the other hand, it is possible to build up a simple picture of the cathode region of a hot-cathode discharge, i.e. one for which the emission of electrons is thermionic.

Another situation which is amenable to treatment by extension of the ideas presented earlier in this chapter is that which occurs when the column has a sudden change in cross-section, or, otherwise, two plasmas of different parameters are adjacent to one another. The plasma balance equation (eqn (2.4)) effectively relates the electron temperature to the similarity variable radius multiplied by pressure, and thus a change in radius implies a change in the electron temperature of the plasma. Also, with a constant total axial current, there will be a change in current density and number density. This implies that there are effectively two different plasma regions adjacent to one another and equilibrium can be maintained only if there is a potential difference between the two regulating the flow of charged particles. This region is one of space-charge separation and is effectively a sheath or boundary layer to each, and hence is referred to as a *double sheath*.

4.9. The cathode region of a hot-cathode discharge

Consider the following model in plane geometry. The cathode is a plane emitting an electron beam of current density J_b, the electrons being accelerated through a space-charge region to the plasma of the positive column. From this plasma ions are accelerated to the cathode while the electrons are in a decelerating field and are in the main returned to the plasma as in a wall sheath.

Suppose the plasma has a density of electrons n_0 without the contribution from the beam and that in the plasma the beam density is $\alpha_b n_0$, then the ion density in the plasma is $(1+\alpha_b)n_0$ for equality of charge density. From continuity there must be a current of ions in the plasma which implies that they must have a drift velocity v_0 and their motion will be given by $\frac{1}{2}M\dot{v}_i^2 = \frac{1}{2}Mv_0^2 - eV$, while continuity requires $n_i v_i = n_0 v_0(1+\alpha_b)$. The 'plasma' electrons will have a density given by the Boltzmann relation $n_p = n_0 \exp(eV/k_B T_e)$, while for the beam we have $\frac{1}{2}mv_b^2 = -e(V_c - V)$ and $n_b e v_b = J_b$, where v_b and n_b are the electron beam velocity and density at any point. Furthermore, the potential is given by the space charge

$$\frac{d^2 V}{dx^2} = -\frac{e(n_i - n_e)}{\varepsilon_0},$$

with $n_e = n_p + n_b$.

The equations can readily be normalized introducing

$$\eta = -\frac{eV}{k_B T_e} \quad \text{and} \quad \xi = \frac{x}{\lambda_D},$$

with the Debye length based on n_0, to give

$$\frac{d^2\eta}{d\xi^2} = (1+\alpha_b)\left(1+2\eta\frac{c_s^2}{v_0^2}\right)^{-\frac{1}{2}} - \exp(-\eta) - \alpha_b \eta_c^{\frac{1}{2}}(\eta_c - \eta)^{-\frac{1}{2}}. \quad (4.17)$$

It can be shown that

$$\alpha_b = \frac{J_b}{n_0 e(2eV_c/m)^{\frac{1}{2}}},$$

since

$$v_b = \left(\frac{2k_B T_e}{m}\right)^{\frac{1}{2}}(\eta_c - \eta)^{\frac{1}{2}} \quad \text{and} \quad n_b = \frac{J_b}{e(2k_B T_e/m)^{\frac{1}{2}}}(\eta_c - \eta)^{-\frac{1}{2}}.$$

By imposing equality of charge density as $\eta \to 0$ we have ensured one of the matching conditions as ξ becomes large. Also if η is to go smoothly to zero as ξ increases, then it may be supposed to vary asymptotically as $K\xi^{-n}$.

Substitution in (4.17) gives

$$n(n+1)K\xi^{-(n+2)} \sim (1+\alpha_b)\left(1 - K\xi^{-n}\frac{c_s^2}{v_0^2} + \frac{3}{2}K^2\xi^{-2n}\frac{c_s^4}{v_0^4}\right) -$$

$$- \left(1 - K\xi^{-n} + \frac{K^2}{2}\xi^{-2n}\right) - \alpha_b\left(1 + \frac{1}{2\eta_c}K\xi^{-n} + \frac{3}{8\eta_c^2}K^2\xi^{-2n}\right),$$

which can be satisfied only if the terms in ξ^{-n} on the right-hand side vanish, i.e.

$$(1+\alpha_b)\frac{c_s^2}{v_0^2} = 1 - \frac{\alpha_b}{2\eta_c}$$

or

$$v_0^2 = c_s^2(1+\alpha_b) \Big/ \left(1 - \frac{\alpha_b}{2\eta_c}\right),$$

and then K can be found from the next-highest-order terms, provided $n = 2$. This condition on the ion velocity is the same as the effective ion acoustic speed in a beam plasma configuration given by Pak and Emeleus (1971).

The field can be found by straightforward integration of (4.17) and is given by

$$\frac{1}{2}\left(\frac{d\eta}{d\xi}\right)^2 = \frac{(1+\alpha_b)^2}{1-(\alpha_b/2\eta_c)}\left(\left\{1+\frac{2\eta\{1-(\alpha_b/2\eta_c)\}\}^{\frac{1}{2}}}{1+\alpha_b}\right\}^{\frac{1}{2}}-1\right) - $$
$$-1+\exp(-\eta)+2\alpha_b\eta_c^{\frac{1}{2}}\{(\eta_c-\eta)^{\frac{1}{2}}-\eta_c^{\frac{1}{2}}\}, \tag{4.18}$$

using $d\eta/d\xi \to 0$ as $\eta \to 0$, and the field at the cathode ε_c is found by putting $\eta = \eta_c$.

The beam current cannot be increased indefinitely for a given η_c, and increase in the cathode temperature merely causes space-charge limitation of the current. Under these circumstances there is a 'virtual cathode' where the field goes to zero a short distance in front of the real cathode (see e.g. Langmuir 1920). The distance is small on the total sheath scale and the potential depression small for η_c large compared with 1. Thus, in the maximum-beam-current situation, it is reasonable to take $\varepsilon_c = 0$. Now

$$\tfrac{1}{2}\varepsilon_c^2 = \frac{(1+\alpha_b)^2}{1-(\alpha_b/2\eta_c)}\left\{\left(1+\frac{2\eta_c-\alpha_b}{1+\alpha_b}\right)^{\frac{1}{2}}-1\right\}-1+\exp(-\eta)-2\alpha_b\eta_c,$$

which for $\eta_c \gg 1$ and $\alpha_b \ll 1$ can be approximated by $\alpha_b \sim 1/(2\eta_c)^{\frac{1}{2}}$. Generally $\tfrac{1}{2}\varepsilon_c^2 \sim (2\eta_c)^{\frac{1}{2}}-2-2\alpha_b\eta_c$, or for fixed η_c, ε_c has a parabolic variation with α_b.

The ratio of beam current to random current under space-charge-limited conditions is then

$$\frac{4J_b}{n_0 e\bar{c}_e} = \frac{J_0}{n_0 e}\left(\frac{2\pi m}{k_B T_e}\right)^{\frac{1}{2}} = \alpha_b(2\eta_c)^{\frac{1}{2}}(2\pi)^{\frac{1}{2}} \sim (2\pi)^{\frac{1}{2}}. \tag{4.19}$$

Figs 4.15(a) and (b) show the results of Prewett and Allen (1976) from numerical integration of (4.17) and (4.18), c.f. the analytic approximations derived above. Comparison is made in Fig. 4.16 with the measurements of

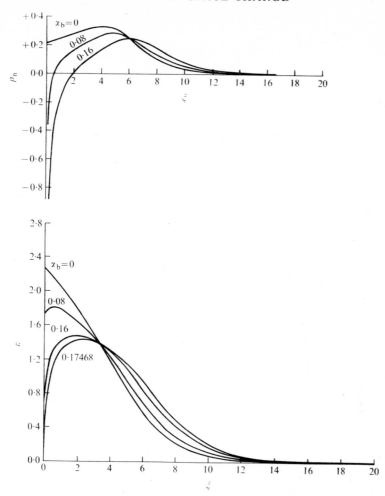

Fig. 4.15. (a) The space-charge density ρ_n near a hot emitting cathode (normalized by $n_0 e$, where n_0 is the plasma density) as a function of distance normalized to the plasma Debye length ξ for different values of the cathode-beam current density J_b expressed in terms of $\alpha_b = J_b/[n_0 e(2eV_c/m)^{\frac{1}{2}}]$ for $eV_c/k_B T_e = 10$. (b) Corresponding distributions of the normalized field ε where $\alpha_b = 0.17468$ corresponds to $\varepsilon(0) = 0$.

Crawford and Cannara (1965), which were obtained by using a probing beam transverse to the discharge to measure the space-charge fields.

The above procedure for joining plasma and sheath has not been carried out in as great a detail as that for the wall sheath but could be if some model were adopted for generation and loss processes axially within the column.

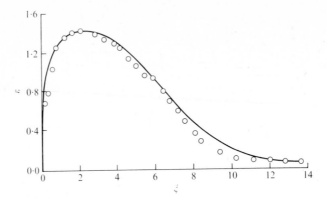

Fɪɢ. 4.16. The measured electric field near a hot cathode compared with theory for zero field at the cathode, parameters being chosen to fit at the peak field.

4.10. The double sheath

The equipotentials, field lines, and particle trajectories which exist near a constriction in discharge cross-section are shown in Fig. 4.17.

The transition region is expected and observed to be of the order of several Debye lengths thick, and since this is typically two or three orders of magnitude less than the discharge radius it is reasonable to adopt a plane model. A schematic sketch of the potential field and space-charge variation with smooth joining to different plasmas is also shown in Fig.

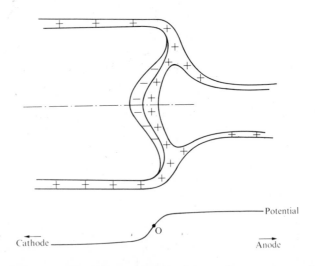

Fɪɢ. 4.17. A schematic diagram of the regions of space charge near a constriction in a discharge indicating regions of net positive charge by + and negative by −. Also shown is the potential along the centre line.

4.17. Similar situations occur in the solid and liquid states and are referred to as double layers, finding particular application because the charge separation represents a capacity which can be varied by varying the voltage across it (Shockley 1949). Also similar regions occur at the boundary grids in double plasma devices (Taylor, Baker, and Ikezi 1970). Let us refer to cathode and anode sides of the layer and the plasmas there as P_c and P_a. Then in the double layer there will be charged particles of both signs originating from both plasmas.

As with the cathode sheath we will assume that those charged particles repelled by the field are given by the appropriate Boltzmann factor; thus from P_c there is an ion density $n_i = n_{ic} \exp(-eV/k_B T_{ic})$ and from P_a an electron density $n_e = n_{ea} \exp\{-e(V_s - V)/k_B T_{ea}\}$ where T_{ic} is the cathode plasma ion temperature and T_{ea} is the anode plasma electron temperature. The particles of opposite sign will be accelerated and give the current flow through the layer so that using continuity we can write the electron density from P_c as $n_e = n_{ec}\{1 + (2eV/mv_{e0}^2)^{-\frac{1}{2}}$, where v_{e0} is the drift velocity of electrons leaving P_c and correspondingly the ion density from P_a as

$$n_i = n_{ia}\left\{1 + \frac{2e(V_s - V)}{Mv_{i0}^2}\right\}^{-\frac{1}{2}}.$$

The boundary conditions to be applied are equality of charge density as $V \to 0$ and V_s and $dV/dx \to 0$ as $V \to 0$ and V_s.

Let us choose as origin the point of zero space-charge density (O of Fig. 4.17) and introduce a Debye length defined by a density n_0 and temperature T_{ea}, which we may assume to be the maximum of the two plasma densities, and normalize lengths to it so that $\xi = x/\lambda_D$.

Then we may take $V_s - V \sim A/\xi^n$ and $V \sim C/\xi^n$ as $\xi \to +\infty$ and $-\infty$ respectively and the boundary conditions give, in a similar manner to previously, four equations in the normalized densities n_{ec}, n_{ic}, n_{ea}, n_{ia}, being n_{ec}/n_0, n_{ic}/n_0, n_{ea}/n_0, and n_{ia}/n_0 respectively. Introducing the parameters

$$\phi_e = \frac{mv_{e0}^2}{2eV_s}, \qquad \phi_i = \frac{Mv_{i0}^2}{2eV_s}, \qquad \tau_e = \frac{k_B T_{ea}}{eV_s}, \quad \text{and} \quad \tau_i = \frac{k_B T_{ic}}{eV_s},$$

the equations become

$$n'_{ic} + \frac{n'_{ia}}{(1 + 1/\phi_i)^{\frac{1}{2}}} = n'_{ec} + n'_{ea} \exp(-1/\tau_e), \tag{4.20}$$

$$n'_{ic} \exp(-1/\tau_i) + n'_{ia} = \frac{n'_{ec}}{(1 + 1/\phi_e)^{\frac{1}{2}}} + n'_{ea}, \tag{4.21}$$

$$-\frac{n'_{ic}}{\tau_i} + \frac{n'_{ia}}{2\phi_i(1 + 1/\phi_i)^{\frac{3}{2}}} = -\frac{n'_{ec}}{2\phi_e} + \frac{n'_{ea}}{\tau_e} \exp(-1/\tau_e), \tag{4.22}$$

and

$$\frac{n'_{ic}}{\tau_i}\exp(-1/\tau_i) - \frac{n'_{ia}}{2\phi_i} = \frac{n'_{ec}}{2\phi_e(1+1/\phi_e)^{\frac{3}{2}}} - \frac{n'_{ea}}{\tau_e}. \qquad (4.23)$$

The determinant of these four homogeneous equations relates ϕ_e, ϕ_i, τ_e, and τ_i.

It can be shown that solutions exist only for $\phi_i \gtrsim \tau_e/2$ and $\phi_e \gtrsim \tau_i/2$, and these are the Bohm criteria for the two boundaries (Andrews 1969). For simplicity and because it is physically close to the truth, we will consider $\tau_i \to 0$, in which case the term in n'_{ic} vanishes from (4.21), (4.22), and (4.23).

The relation between ϕ_e, ϕ_i, and τ_e which results is shown in Fig. 4.18.

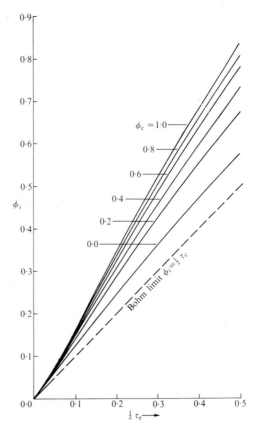

FIG. 4.18. The relation between the normalized directed ion energy $\phi_i = Mv_{i0}^2/2eV_s$, where V_s is the potential across a double sheath and the normalized electron temperature of the anode plasma $\tau_e = k_B T_{ea}/eV_s$ with the normalized directed electron energy $\phi_e = mv_{eo}^2/2eV_s$ as a parameter. The Bohm criterion for a single sheath is also shown.

For any set of values then the ratios of n'_{ia}, n'_{ea}, and n'_{ec} can be found and substitution in (4.20) gives n'_{ic}.

In order to calculate spatial variations we need to relate V to x, i.e. to locate the origin O. Denoting the value of V at O as V_0, it is given by

$$n'_{ia}\left\{1+\frac{2e(V_s-V_0)}{Mv_{i0}^2}\right\}^{-\frac{1}{2}} = n'_{ec}\left(1+\frac{2eV_0}{mv_{e0}^2}\right)^{-\frac{1}{2}} + n'_{ea}\exp\left\{-\frac{e(V_s-V_0)}{k_B T_{ea}}\right\}$$

or

$$n'_{ia}\left(1+\frac{1-\phi_0}{\phi_i}\right)^{-\frac{1}{2}} = n'_{ec}\left(1+\frac{\phi_0}{\phi_e}\right)^{-\frac{1}{2}} + n'_{ea}\exp\left(\frac{\phi_0-1}{\tau_e}\right),$$

where

$$\phi_0 = V_0/V_s.$$

Integration from the origin can proceed if we know the field there. This can be found by integration from either $+\infty$ or $-\infty$ using the fact that $dV/dx \to 0$ as $V \to 0$. Thus integrating from $+\infty$

$$-\tfrac{1}{2}\varepsilon_0\left(\frac{dV}{dx}\right)_0^2 = n_{ia}Mv_{ia}^2\left[\left\{1+\frac{2e(V_s-V_0)}{Mv_{i0}^2}\right\}^2 - 1\right] -$$

$$- n_{ec}mv_{e0}^2\left\{\left(1+\frac{2eV_s}{mv_{e0}^2}\right)^{\frac{1}{2}} - \left(1+\frac{2eV_0}{mv_{e0}^2}\right)^{\frac{1}{2}}\right\} -$$

$$- n_{ea}k_B T_{ea}\left[1-\exp\left\{-\frac{e(V_s-V_0)}{k_B T_{ea}}\right\}\right]\right]$$

or

$$\tau_e\left(\frac{d\phi}{d\xi}\right)_0^2 = -4n'_{ia}\phi_i\left\{\left(1+\frac{1-\phi_0}{\phi_i}\right)^{\frac{1}{2}} - 1\right\} +$$

$$+ 4n'_{ec}\phi_e\left\{\left(1+\frac{1}{\phi_e}\right)^{\frac{1}{2}} - \left(1+\frac{\phi_0}{\phi_e}\right)^{\frac{1}{2}}\right\} + 2n'_{ea}\tau_e\left\{1-\exp\left(\frac{\phi_0-1}{\tau_e}\right)\right\}.$$

Numerical integration can now proceed and a typical solution is shown in Fig. 4.19, where the results of Andrews (1969) demonstrate the double space-charge layer with ρ going smoothly to zero at $\eta = 1$ (since $T_e \neq 0$) but going abruptly to zero at $\eta = 0$ (since $T_i = 0$). Also the set of equations used by him included one stating

$$\int_0^{V_s} \rho\, dV = 0;$$

but since

$$\int_0^{V_s} \rho\, dV = \int_{-\infty}^{\infty} \frac{d^2V}{dx^2}\cdot\frac{dV}{dx} = \frac{1}{2}\left(\frac{dV}{dx}\right)_{+\infty}^2 - \frac{1}{2}\left(\frac{dV}{dx}\right)_{-\infty}^2$$

it is equivalent to a vanishing field condition. Computations have been carried through for $T_i \neq 0$, and for details of these and a further discussion

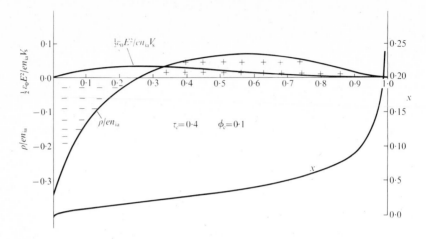

Fɪɢ. 4.19. The space charge, electrostatic energy density, and distance as functions of the potential through a double sheath for parameters $\tau_e = 0\cdot4$, $\phi_e = 0\cdot1$. The space charge ρ is normalized to the anode plasma ion density n_{ia}, the energy density to $n_{ia}eV_s$, and distance to $(\varepsilon_0 V_s/2n_{ia}e)^{\frac{1}{2}}$, a modified Debye length.

of theoretical aspects of the double sheath the reader is referred to Andrews and Allen (1971).

A further stage in the development of the theory of the double sheath would be the relating of the double-sheath voltage and current, $A(n_{ec}v_{e0} + n_{ia}v_{i0})$ to the parameters of the plasmas bounding it.

The applications that have been made of the treatment above do not depend on the details of the structure of the double sheath. No measurements like those given for the hot-cathode sheath are available for comparison with theory, but the phenomenon of current limitation in constricted discharges seems to be well accounted for if ion fluxes and hence particle loss rates are calculated on the basis of the above model (Andrews 1969).

The fact that an almost monoenergetic beam of electrons emerges from the cathode plasma has been proposed as a method of carrying out fundamental measurements with low-energy beams which are otherwise experimentally difficult to obtain (Andrews and Varey 1970).

5

RADIALLY VARYING AND TIME VARYING POSITIVE COLUMNS

5.1. Variation of neutral density

IN EARLIER chapters we have assumed that variations in the neutral gas density and temperature across the column were negligible. There are circumstances in which this is no longer true and the consequences of heating of the gas by the ohmic heating of the discharge current will be considered in later sections of this chapter. In this section we consider two low-pressure phenomena which limit the validity of earlier treatments.

The first arises from the elevated electron temperatures in low-pressure discharges. Since T_e increases as pr_w decreases the situation can arise where, for a given electron density, the electron partial pressure $n_e k_B T_e$ becomes significant in comparison with the neutral pressure $n_g k_B T_g$. The generation terms then in the plasma balance equation must be modified and so the generation rate for single-stage processes becomes

$$n_e(r)n_g(r)\frac{Z}{n_{g0}}(T_e),$$

where, since the total pressure p is constant,

$$p = n_e(r)k_B T_e + n_g(r)k_B T_g,$$

and thus

$$n_g = \frac{p - n_e k_B T_e}{k_B T_g}.$$

Other physical quantities, e.g. the light output, will also be affected. Ion pressure effects will be of comparable importance only in situations where $T_i \sim T_e$, such as in arcs at high pressure.

This partial-pressure effect was studied by Steenbeck (1939), who detected differences in the neutral gas pressure between the centre and wall in an argon discharge at 40 mTorr. Fig. 5.1 shows a comparison between this pressure difference and $n_e(0)k_B T_e$ measured by electrostatic probes as a function of current. At somewhat lower pressures, namely 3–30 mTorr, Franklin (1963) measured the radial distribution of the light output from a mercury discharge for which light emission was measured

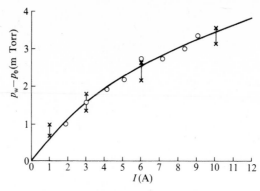

FIG. 5.1. A comparison between the pressure difference measured between the centre of the discharge and the wall, $p_w - p_0$, as a function of discharge current I and the electron pressure $n_{e0} k_B T_e$, $\times$ and theory for argon gas at 40 mTorr.

along a chord and compared with $\int n_e(r) n_g(r)\, dx$, electron number densities being determined from microwave properties, with the conclusion that the ratio

$$\alpha_p = \frac{n_e(0) k_B T_e}{n_e(r) k_B T_e + n_g(r) k_B T_e} \equiv \frac{n_e(0) k_B T_e}{p}$$

reached a value 0·8 for the highest currents observed. A comparison between theory and experiment is given in Figs 5.2(a) and (b). Recent measurements at pressures of 200–500 mTorr in argon lasers have been made by Ebert (1975).

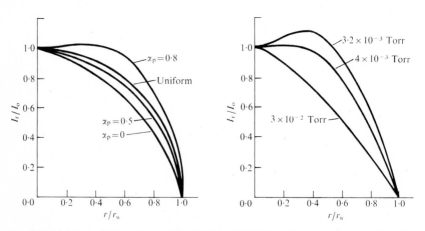

FIG. 5.2. (a) The variation of normalized light output I_r/I_0 along a chord as a function of the parameter $\alpha_p = n_{e0} k_B T_e / p_w$. For comparison that from a radially uniform discharge is also shown. (b) The measured light output for a mercury discharge tube of radius 4·7 mm at pressures of 30 mTorr, 4 mTorr, and 3·2 mTorr and number densities corresponding to values of α_p of 0·085, 0·64, and 0·8 respectively.

A second process which affects the neutral density distribution in low-pressure discharge is due to the radial motion of ions. While an atom is ionized it is unable to contribute to the process of generation of further charges, it being assumed that there is a negligible small number of electrons capable of doubly ionizing a gas atom. It is also true that neutralization of the ions takes place principally on the wall of the discharge vessel which, if an insulator, may require a significant time for an electron to become available at the place required. This wall return effect is responsible for hysteresis observed at low modulation frequencies in mercury and has been carefully studied by Riley (1970), using a model proposed by Kenty (1938) in which the atom wall density Q is determined by the wall-directed ion flux Γ_{ir}, the absorption coefficient A, for ions striking the wall, the neutral emission per ion BQ due to ion impact with the wall, and the rate Q/τ of spontaneous emission of neutrals, so that

$$\frac{dQ}{dt} = A\Gamma_{ir}Q - \frac{Q}{\tau}.$$

Under active discharge conditions the neutral flux to the wall is negligible in comparison with the ion flux.

This picture of ions arriving at the wall and returning to the body of the discharge as neutrals allows a model to be set up to explain a feature of low-pressure discharges. This is what is often referred to as current limitation or the extinction of a discharge if an attempt is made to pass a current outside defined limits. A lower bound arises (as indicated in Chapter 4) from space-charge effects, since as λ_D/r_w increases the value of $Zr_wM^{\frac{1}{2}}/(k_BT_e)^{\frac{1}{2}}$ increases rapidly. This requires an increase in axial field larger than can be supplied, or if such a voltage-source limitation does not exist the reduced longitudinal field rises until E_z/p exceeds that corresponding to the maximum of $Z/p(E_z/p)$ or, equivalently, $\dot{Z}/p(T_e)$. If we designate this value of T_e as T_e^* and the corresponding Z/p as Z^*/p then

$$\frac{\xi_w}{pr_wM^{\frac{1}{2}}} \lesssim \left(\frac{Z}{p}\right)^* \frac{1}{(k_BT_e^*)^{\frac{1}{2}}}$$

or

$$\xi_w^* = pr_wM^{\frac{1}{2}}C_1,$$

implying

$$\alpha^2 \lesssim \alpha^2(\xi_w^*)$$

or

$$I \gtrsim \varepsilon_0 \frac{(k_BT_e^*)^{\frac{3}{2}}}{\alpha^2(\xi_w^*)} C_2 \equiv I_{LL},$$

setting $v_D = C_2'(k_BT_e^*)^{\frac{1}{2}}$. I_{LL}, the lower limit to the current, may be expected to decrease as pr_w increases.

The wall return model of the mercury discharge allows an upper bound to the current to be determined, for if the incoming ion flux exceeds the outgoing neutral flux, there will be a depletion of neutral gas density in the volume. This will occur for

$$A\Gamma_{\text{ir}} = BQ\Gamma_{\text{ir}} + \frac{Q}{\tau}$$

or

$$\Gamma_{\text{ir}} = \frac{Q}{\tau(A - BQ)}.$$

Denoting the right-hand side ν_0 and a constant for a given wall temperature, and setting $v_D = C_2'(k_B T_e)^{\frac{1}{2}}$ and $\Gamma = n_{e0} C_3 r_w (k_B T_e / M)^{\frac{1}{2}}$, as shown in (4.16), leads to

$$I_{\text{UL}} = n_{e0} r_w^2 C_4 v_D = \nu_0 r_w C_2' C_4 M^{\frac{1}{2}} / C_3,$$

or $I_{\text{UL}} \propto \nu_0 r_w$, where I_{UL} is the upper limit to the current. This picture needs to be corrected to take into account the finite mean free path of neutrals with respect to ionization which causes I_{UL} to decrease at low pressures and leads to a limiting value of the similarity variable pr_w below which a glow discharge cannot be initiated. Using an exact analysis of the neutral flux at any radial position to calculate number density and current, Stangeby and Allen (1971) estimate this to be $pr_w = 0 \cdot 1$ mm mTorr for mercury, and their high and low current limits can be combined to give Fig. 5.3, where comparison is made with experimental values of the current below which it was not possible to maintain a discharge. It can be seen that the upper and lower limit curves cross and that there is a lower limit, in terms of the variables, pressure, radius, and temperature, below which no low-pressure mercury positive column exist.

The model given by Stangeby and Allen described above was based on the results of the treatment of the uniform column given in Chapter 2, with the plasma balance equation modified for the finite mean free path for ionization. Valentini (1972, 1974) has incorporated this feature into the free-fall theory of the positive column. He has taken the result derived by Caruso and Cavaliere (1964) giving the neutral atom density at a distance x from the midplane of a plane discharge when neutral atoms are derived from neutralization at the wall and leave the wall with a normal velocity component v_{nw}. In differential form the neutral density is given by

$$(n_n^2(x) - n_{n0}^2)^{-\frac{1}{2}} \frac{d n_n(x)}{dx} = \frac{n_e(x) Z_0}{v_{\text{nw}}}, \tag{5.1}$$

where Z_0 is the ionization rate $Z(T_e)$ per unit central density n_{n0}, and thus is a rate constant for ionization. This equation can then be solved

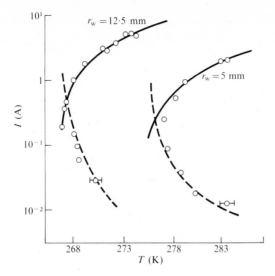

FIG. 5.3. The limiting current I in a mercury discharge as a function of the reservoir temperature T for discharges of radius $r_w = 12.5$ mm and 5 mm compared with theoretical values for the upper limit (solid line) and lower limit (dashed line).

together with those for the charged particles in either the free fall or fluids models of Chapter 2, taking account of the variation of ionization rate with gas density.

From the form of (5.1) it is readily shown that

$$\Lambda = \frac{n_{e0}}{n_{n0}} \cdot \frac{c_s}{v_{nw}} \tag{5.2}$$

is a parameter of the problem. Valentini (1972) has solved the free-fall case showing that the plasma boundary occurs at the same value of the 'wall potential' $\eta_w \equiv 0.8539$. The normalized wall position ξ_w can thus be found numerically and expressed as

$$\xi_w = \xi_{w0} F_1(\Lambda) = \xi_{w0} G_1(d/\lambda_{in}),$$

where d/λ_{in} is the ratio of the lateral dimension to the mean free path for ionization of neutrals since $d/\lambda_{in} \equiv \Lambda \xi_w$. ξ_w as a function of Λ is given in Table 5.1.

TABLE 5.1

Λ	0	0·1	0·2	0·5	0·8	1·0	1·2	1·3	1·4	1·45	1·4515
ξ_w	0·5722	0·593	0·615	0·700	0·829	0·962	1·192	1·406	1·88	3·32	∞

Valentini (1974) also derived the fractional ionization as a function of position regarding v_{n0}/c_s as a parameter which can be related to n_{e0}/n_{n0}

using eqn (5.2). Current limitation then appears as

$$\frac{n_{e0}}{n_{e0}+n_{n0}} \to 1,$$

i.e. the fractional ionization tends to unity or the neutral density tends to zero at the centre of the discharge. Detail comparison with experiment has yet to be made.

5.2. Radial variation of the charged-particle temperatures

So far in the treatment of the positive column the particle energy equations have not been involved. It has been assumed that the electron and ion temperatures are radially uniform and that for a given gas there is a relation between the reduced longitudinal field E_z/p and the electron temperature which determines the field necessary to balance generation and loss processes.

By using the energy equations it is possible to explore this relationship, and in its simplest form it would require

$$E_z I_z = \frac{T_e - T_n}{\tau_T} \int_0^{r_w} n_e 2\pi r \, dr,$$

where τ_T is the electron thermal relaxation time.

Ilic has carried through a much more general calculation in which T_e and T_i are allowed to vary radially and the energy loss involved in ionization is taken into account (Ilic 1973). Thermal conduction is neglected so that the energy equation becomes

$$\mathbf{v} \cdot \nabla(\tfrac{1}{2}m_\alpha v^2 + \tfrac{5}{2}k_B T_\alpha) + Z(\tfrac{1}{2}m_\alpha v^2 + \tfrac{5}{2}k_B T_\alpha) = e\mathbf{v}_\alpha \cdot \mathbf{E} + \mathscr{I}_\alpha$$

where for electrons

$$\mathscr{I} = -ZeV_i - \frac{3m}{M}k_B T_e \nu_e$$

and for ions

$$\mathscr{I} = \tfrac{3}{2}Zk_B T_n - \tfrac{3}{2}\nu_i k_B(T_i - T_n).$$

The model then takes into account ionization and elastic collisions but ignores the effect of excitation for electrons, while for ions there is an input due to ionization and a loss due to charge-exchange collisions. Ion energies do not become sufficiently large for them to participate in the generation process by ion–atom impact. The electron velocity distribution is generalized to take into account the axial drift of electrons and thus the ionization rate becomes a function of the drift velocity v_z which is radially uniform and a magnetic field is included in the momentum equations so that azimuthal components of velocity occur. Guided by the results of

Chapter 4 the charged-particle densities are set equal and the limitations noted.

There results seven first-order differential equations for the variables N, u, $u_{e\theta}$, $u_{i\theta}$, T_e, T_i, and the potential ϕ, which, given appropriate boundary conditions, can be integrated to find them as a function of radius. The equations have features in common with the more restricted treatments:

(1) the singularity at the Bohm speed persists and determines the outer boundary but is now generalized to

$$v^2 = \frac{5}{3}\frac{k_B\{T_e(r_w) + T_i(r_w)\}}{M + m};$$

(2) setting the radial derivatives of temperature, density, and potential to zero on the axis requires the central electron and ion temperatures to be related to the longitudinal field and the ionization rate, and comparison can be made with experimentally determined quantities, as is shown in Fig. 5.4. $k_B T_{e0}$ is given approximately by

$$\frac{Z + 2\nu_e}{(Z + \nu_e)^2}\frac{e^2 E_z^2}{5mZ} - \tfrac{2}{5}eV_i.$$

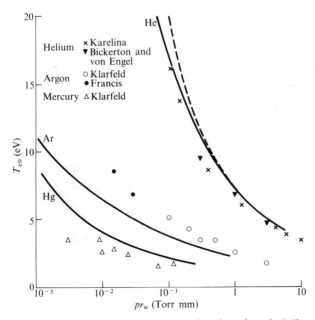

FIG. 5.4. The central electron temperature T_{e0} as a function of pr_w in helium, argon, and mercury as calculated theoretically compared with experimental measurements. The effect of electron drift is shown for helium.

The detailed radial variations of the radial velocity, potential, and number density are given in Fig. 5.5, and are seen to be closely similar to those for parameters corresponding to helium over a range of pressures found in the simpler treatments. The new information shows that there is some variation of both electron and ion temperatures. Examination of the momentum equations for ions and electrons shows that the radial reduction in electron temperature matches the increase in ion temperature plus the ion kinetic (or radial drift) energy together with gas heating by collisions. The assumption of a uniform electron temperature which has been the basis of earlier chapters is seen to be best justified at higher pressures, when the variation is confined to a region near the wall. The ion temperature is seen to increase significantly radially due to the radial motion of ions becoming randomized. The increase is not strongly dependent on pressure. However, as the pressure increases the ions and gas atoms come into good thermal contact and the temperatures implied by $T_i/T_{n0} \sim 40$ (some 12 000 K) would not be physically possible with normal (glass) walls, implying that some improvement is needed in the model for the ion energy equation if physically reasonable answers are to be obtained. This would appear to be the inclusion of thermal conduction.

5.3. The positive column with negative ions

A number of gases, particularly those in groups VI and VII of the periodic table, are electron-attaching and therefore any discharge in such a gas in the steady state will contain a proportion of negative ions. This means that, in setting up the equations for generation and loss, both electron attachment and detachment need to be taken into account and the model of the positive column has to be extended to become a three-fluid model. Supposing the attachment coefficient is A and the detachment coefficient is D and denoting quantities associated with the negative ions by the suffix 'n', the equations become

$$\nabla \cdot (n_e \mathbf{v}_e) = (Z - A)n_e + Dn_n, \quad \text{electron continuity;} \quad (5.3)$$

$$\nabla \cdot (n_i \mathbf{v}_i) = Zn_e, \quad \text{positive-ion continuity;} \quad (5.4)$$

$$\nabla \cdot (n_n \mathbf{v}_n) = An_e - Dn_n, \quad \text{negative-ion continuity;} \quad (5.5)$$

and these give $n_i \mathbf{v}_i = n_e \mathbf{v}_e + n_n \mathbf{v}_n$ when the boundary conditions $\mathbf{v}_i = 0$, $\mathbf{v}_e = 0$, $\mathbf{v}_n = 0$ are imposed at $r = 0$, the centre of the plasma.

The equations of motion for the negatively charged particles taking account of the adverse radial field can be written, ignoring the inertial

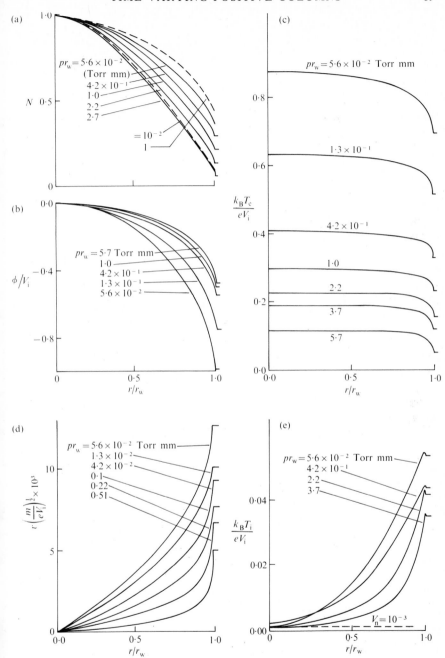

FIG. 5.5. The radial variations of normalized number density N, potential ϕ electron temperature T_e, radial velocity v and ion temperature T_i as a function of the parameter pr_w for helium when T_e and T_i are allowed to vary radially.

term,

$$\frac{k_B T_e}{m} \nabla n_e - \frac{n_e e}{m} \nabla V + \nu_e n_e \mathbf{v}_e = 0, \tag{5.6}$$

$$\frac{k_B T_n}{M_n} \nabla n_n - \frac{n_n e}{M_n} \nabla V + \nu_n n_n \mathbf{v}_n = 0, \tag{5.7}$$

while for positive ions, inertia must be included,

$$+ \frac{n_i e}{M_i} \nabla V + \nu_i n_i \mathbf{v}_i + \nabla \cdot (n_i \mathbf{v}_i \mathbf{v}_i) = 0. \tag{5.8}$$

For each type of particle it has been assumed that the terms arising from volume generation and loss are negligible in comparison with the collisional term, i.e. $\nu_e \gg Z, A$, etc. The set is completed with

$$\mathbf{E} = -\nabla V,$$

$$\nabla \cdot \mathbf{E} = \frac{e}{\varepsilon_0} (n_i - n_n - n_e).$$

Following earlier chapters we normalize number densities to n_{e0}, the central electron density, particle velocities to c_s, the positive ion sound speed, lengths to c_s/Z, and the potential to $-k_B T_e/e$. Appropriate parameters are $\delta_i' = \nu_i/Z$ and $\delta_e' = m\nu_e/M_i Z$ as before, and additionally $\delta_n' = M_n \nu_n/M_i Z$,

$$\varepsilon = T_n/T_e, \qquad \lambda = A/Z, \qquad \mu = D/Z, \quad \text{and} \quad \alpha^2 = \frac{\varepsilon_0 M Z^2}{n_{e0} e^2} = \left(\frac{n_{i0}}{n_{e0}}\right)\left(\frac{Z^2}{\omega_{pi}^2}\right).$$

The resulting seven equations in the unknowns

$$y_e = \frac{n_e}{n_{e0}}, \qquad y_n = \frac{n_n}{n_{e0}}, \qquad u_i = \frac{v_i}{c_s}, \qquad \Gamma_i = \frac{n_i v_i}{n_{e0} c_s},$$

$$\varepsilon = \frac{eE c_s}{k_B T_e Z}, \qquad \eta = \frac{eV}{k_B T_e}, \quad \text{and} \quad \Gamma_n = \frac{n_n v_n}{n_{e0} c_s},$$

(u_e being determined from $y_e u_e = y_i u_i = y_n u_n$) can be solved subject to appropriate boundary conditions. Similarly to the treatment of the case without negative ions, described in Chapter 4, integration can proceed outwards from the centre (having used Taylor-series approximations to achieve stability), but the value of the central ratio of negative-ion density to electron density is an eigenvalue of the problem which is determined by the conditions at the wall.

The results of eqn (4.16) suggested that the wall boundary condition on any diffusing species of particle should be determined by

$$\Gamma_{net} = nv = \frac{n\bar{c}(1-r)}{2(1+r)}.$$

Applying this separately to electrons and negative ions means that both u_e and u_n have predetermined values u_{ew}, u_{nw} at the wall, and it is the simultaneous satisfaction of these two requirements which allows the eigenvalue n_{n0}/n_{e0} to be found. The form of the solutions is such that n_n is vanishingly small relative to n_e near the wall (see Fig. 5.6), and so for numerical reasons the formal condition given above on the negative ion speed at the wall was replaced by the physically more obvious and equivalent one

$$\int_0^{r_w} An_e 2\pi r \, dr - \int_0^{r_w} Dn_n 2\pi r \, dr = 2\pi r_w n_{nw} v_{nw}$$

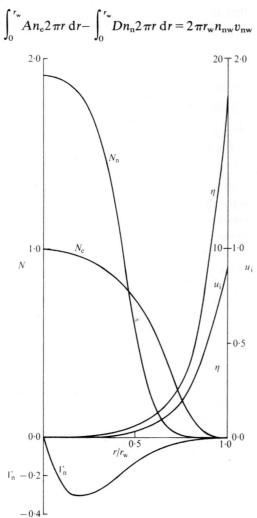

FIG. 5.6. The radial variation relative electron density $N_e = n_e/n_{e0}$, negative ion density $N_n = n_n/n_{e0}$, potential η, ion radial velocity $u_i = v_i/c_s$ and negative ion flux Γ_n for a positive column with negative ions such that $\lambda = 1\cdot0$, $\alpha = 0\cdot003$, $\varepsilon = 0\cdot005$, $\mu = 1\cdot0$, $\delta_i' = 300$, $\delta_e' = 0$, $\delta_n' = 210$.

and the limit $n_n \rightarrow 0$ taken. Thus the negative ion flux at the wall is essentially zero and the negative ions are completely confined to the central regions of the discharge, with the negative ion flux everywhere directed inwards. The 'position' of the wall in the normalized space variable $x = rZ/c_s$ sets the electron temperature, or mean energy, as in the conventional plasma balance equation (2.4).

Having chosen values for the parameters λ, μ, α, δ'_e, δ'_i, δ'_n, and ε then n_{n0}/n_{e0} is varied iteratively until the boundary conditions are simultaneously satisfied. This process is time-consuming but, as the results show, the earlier approximation of setting $n_n/n_e = $ constant (Holm 1932; Seeliger 1949) is not well satisfied and, further, there are several scale lengths in the problem, so that the labour is necessary in order to understand the positive column with negative ions. The above treatment is due to Edgley (1975) and, being based on a clear foundation, allows one to see the approximations in earlier work, particularly that of Sabadil (1973).

Some results typical of those determined by Edgley are given in Fig. 5.6, where the parameter that is varied is the detachment rate. As detachment becomes rare, relative to ionization, while attachment remains comparable, the negative-ion density builds up in the centre of the discharge and can become many times the electron density. This accords with experimental observation in such gases (see e.g. Thompson 1959). The electron density distribution is modified somewhat in comparison with that in non-attaching gases and shows a point of inflection within the discharge. This means that the visible light output, being proportional to the product of electron and excited atom density, is concentrated in the centre of the discharge. This and other mechanisms relevant to the contraction of the positive column are discussed later in this chapter.

The plasma balance equation relating the generation rate and the dimensions can be compared with experiment by relating the generation rate to the longitudinal field. In order to achieve this some assumption has to be made about the form of the electron energy distribution. Using the Druyvesteyn form the process has been carried through for oxygen by Edgley and comparison between theory and experiment has been made.

5.4. Contraction of the positive column

The theoretical and experimental curves of number density presented so far have had a common feature in that the discharge filled the whole tube to the walls and, except when negative ions are present, the curvature of the number-density distribution has been negative everywhere. There are, however, experimental conditions in which these features are not found, and they are usually referred to as contracted columns. Several mechanisms have been put forward to explain contraction as observed. These and their features will be described below, but it is generally true that contraction occurs at higher pressures in all gases, although more strongly in electron-attaching gases. The contraction may be severe, leading to a filamentary discharge or a diffuse one, as observed in the light output, without a sharp boundary.

Fig. 5.7 gives measurements by Heymann (1969) of the radiation temperature at microwave frequencies in neon and argon, where it is seen that there is a clear and relatively abrupt transition at almost constant

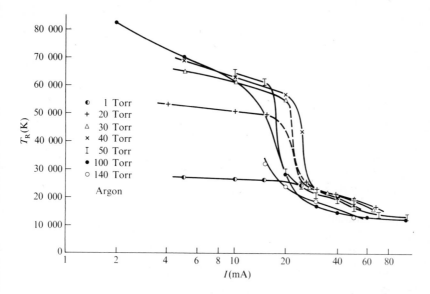

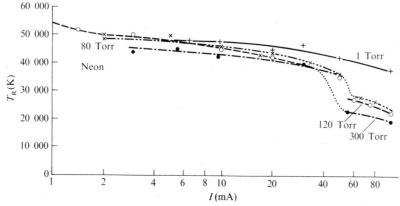

FIG. 5.7. The measured radiation electron temperature T_R as a function of current I at different pressures in a tube of radius $r_w = 13$ mm in (a) argon, (b) neon. The contraction of the positive column is seen as a sudden change in T_R at almost constant current and almost independent of gas pressure.

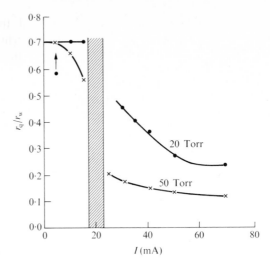

FIG. 5.8. The measured ratio of width at half peak density to diameter for discharges at 20 Torr and 50 Torr in argon gas as a function of discharge current I, showing a sudden change similar to that in Fig. 5.7. The shaded region is one in which the discharge is unstable.

current for pressures greater than a few Torr. Fig. 5.8 presents information on the radial distribution in terms of a parameter defined by the equivalent radius r_q of a uniform discharge with the same peak density as the actual discharge of radius r_w, i.e.

$$n_0 r_q^2 = 2 \int_0^{r_w} n(r) r \, dr.$$

The measurements, made by determining the Q of a microwave cavity as a function of discharge current and pressure, are in quantitative agreement with those in Fig. 5.7.

For the purposes of comparison with later sections it should be noted that the value of the reduced current I/r_w in A mm^{-1} at contraction is ~ 0.002–0.004.

5.5. Thermal effects in the positive column

A simple picture of one possible mechanism can be gained by taking account of gas heating by the discharge current. Because of the radial variation of current density and because of conduction to the walls, one would expect there to be an elevated gas temperature at the centre of the discharge. This would result in a lowering of the density there and thus the effective ionizing field E/n_g would be increased. This in turn would lead to a greater ionization rate and thus a greater current density. Fig. 5.9(a) shows measurements of the radial variation of gas temperature in

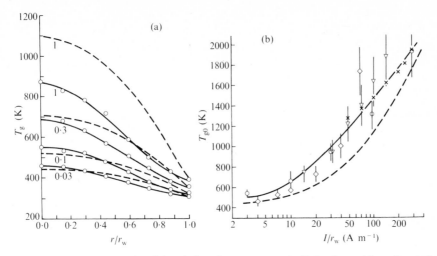

FIG. 5.9. (a) The measured radial variation of gas temperature T_g for the positive column of an argon discharge, $pr_w = 760$ mm Torr as a function of the parameter I/r_w and solid curve compared with the theoretical parabolic profile (dashed). (b) The central gas temperature T_{g0} under the same conditions as a function of I/r_w comparing theory (dashed) and experiment.

argon at relatively high pressures, showing how, as the reduced current I/r_w is increased, the central temperature is raised above the wall temperature. The measurements were made with an iron–constantan thermocouple and are compared with a simple theory based on a parabolic temperature profile and heat loss by thermal conduction alone (Hayess and Wojaczek 1970). The variation of central gas temperature measured under three different sets of conditions and corrected for emission and the contact temperature difference is shown compared with the same theoretical values in Fig. 5.9(b). The values of I/r_w for which gas heating becomes significant are comparable with those at which contraction sets in. However, the above line of reasoning ignores the fact that the particle-loss processes will be enhanced at lower densities, and furthermore these effects cannot change the sign of the curvature of the electron density distribution which will be given by

$$\nabla \cdot (D_a \nabla n_e) + Z n_e = 0.$$

Now ∇n_e and ∇D_a will both be negative since $D_a \propto T_g$, and thus the term $\nabla n_e \cdot \nabla D_a$ cannot allow $\mathrm{d}^2 n_e / \mathrm{d} r^2$ to vanish. Thus simple gas heating alone cannot lead to contraction.

5.6. Volume recombination in the positive column

If, as well as generation by electron impact, we take into account loss by volume recombination between electrons and positive ions in the

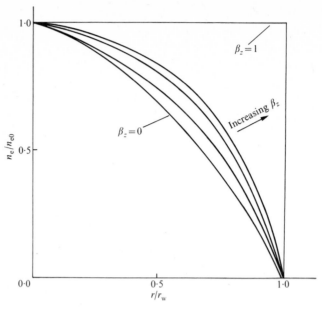

F<small>IG</small>. 5.10. The influence of recombination on the radial distribution of electron density in a discharge where the parameter β_z is the relative value of the recombination and ionization rates at the centre.

presence of a third body, the plasma balance equation would become

$$D_a\nabla^2 n_e + Z n_e - \rho_i n_e^2 = 0.$$

Clearly $d^2 n_e/dr^2$ could vanish now for some appropriate n_e, but in regions of larger n_e, $\nabla^2 n_e$ would be positive, and in regions of smaller n_e it would be negative. Fig. 5.10 shows calculations of the number-density distribution as a function of the parameter $\beta_z \equiv \rho_i n_{e0}/Z$, showing that volume recombination alone causes spreading rather than contraction.

5.7. Contraction due to the combined effects of gas heating and volume recombination

If the theory is taken one stage further by including the effects of variation of gas temperature on recombination coefficient as well as ionization and diffusion, a different picture can emerge. This is due to the fact that the recombination coefficient ρ_i varies more strongly with gas temperature than the ionization coefficient; typically it is found that ρ_i varies as $T_g^{-\frac{3}{2}}$ and the mobility μ_i as $T_g^{-\frac{1}{2}}$. This situation has been considered by Lynch (1967).

It is possible to show from the fact that the energy input per electron per atom is constant across the cross-section that the electron and gas

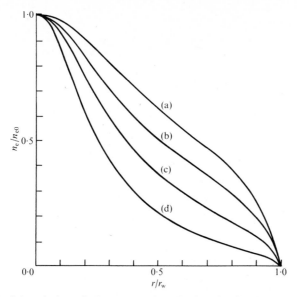

Fig. 5.11. Radial variation of electron number density showing how the temperature variation of the recombination coefficient can cause relative contraction. The parameters for the curves are the reduced variables E_z/p and pr_w, in $V\,Torr^{-1}m^{-1}$ and $m\,Torr$ being respectively, (a) 20, 1·49; (b) 12, 114; (c) 6, 110; (d) 4, 285.

temperatures must be proportional. This means that the ionization rate, which is approximately proportional to $\exp(-eV_i/k_B T_e)$, will vary across the section, being largest in the centre, and it is found that under circumstances corresponding to typical gases, concavity in the number-density distribution can occur, as is seen in Fig. 5.11.

The light output is likely to be a function of electron and excited-state density or electron and gas-atom density, depending on the process involved; the former would produce a volume emission proportional to n_e^2/T_g and the latter proportional to n_e/T_g. This situation is of course such that similarity is not obeyed and increasing the current density increases the magnitude of the contraction.

5.8. Contraction due to radial variation in the electron energy distribution function

In an earlier section of this chapter (§ 5.2) the possibility of electron temperature being a function of radius was allowed in the energy equations and a small variation was found to occur. However, it was assumed, and indeed this is a satisfactory approximation from the point of view of the energy equations, that the distribution function was Maxwellian. When one seeks to characterize the ionization rate it is the high-energy tail of the distribution which is important and not the mean energy. In

Chapter 1 an expression was given which indicated that at low charged-particle densities the high-energy tail would be expected to decrease more rapidly than Maxwellian, and indeed measurements do confirm this under some conditions in some gases.

This feature has been taken into account by Wojaczek (1968) and by Kagan and Lyaguschenko (1962, 1964) who have developed a theory of the contraction which, while it can be extended to include the effects of gas heating and volume recombination, assumes that the ionization and the light output arise from stepwise processes, i.e. collisions of electrons with excited states. (see Appendix I.)

The distribution function for number densities between certain values, which are determined by the nature of the gas, is dependent on the number density and will lie between the Druyvesteyn and Maxwellian limits. With the form of the function having greatest deviation from Maxwellian above the first excitation potential. A detailed calculation in this region suggests a generation rate of the form $Z = K \exp(-Cn_{e0}/n_e)$, where C is a function of the gas and its excitation cross-section. With this form the equation for the radial distribution of number density takes the form

$$D_a \nabla^2 n_e + n_e K \exp\left(-\frac{Cn_{e0}}{n_e}\right) = 0, \qquad (5.9)$$

and the light output may be taken to be proportional to $n_e^2 \exp(-Cn_{e0}/n_e)$. Distributions of electron density and ionization density (approximately light output) are shown in Fig. 5.12(a) for $C = 1$. The fractional width at half-height H, i.e. corresponding to $n_e/n_{e0} = 0.5$, is shown as a function of C in Fig. 5.12(b). C decreases as n_{e0} increases, so

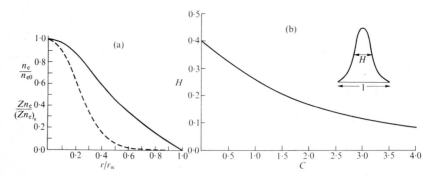

FIG. 5.12. (a) Radial variation of the relative electron density n_e/n_{e0} (solid line) and light output $Zn_e/(Zn_e)_0$ (dashed line) showing the effect of radial variation of the electron distribution function. The curves are for the parameter $C \equiv 1$, where $Z(r) \propto \exp(-Cn_{e0}/n_e)$. (b) Variation of the half-width H as a function of the parameter C showing increased contraction as C increases.

that $n_{eo}C$ is approximately constant. Thus, at constant pr_w, as the current is increased H increases and the contraction is predicted to become less pronounced.

The model can readily be extended to include the effect of recombination, giving an equation

$$D_a \nabla^2 n_e + K n_e \exp(-C n_{eo}/n_e) - \rho_i n_e^2 = 0.$$

Defining $\beta_Z = (\rho_i n_e/Z)_0$ as earlier in this chapter gives results which show greater contraction as β_Z increases until a critical value of 0·4 is reached. Wojaczek's results for the radial distribution are given in Fig. 5.13 for $C = 1$ as β_Z is varied, while Fig. 5.14 compares experimental and theoretical curves for the half-width in argon for $pr_w = 500$ mm Torr as a function of the reduced current I/r_w for different values of the recombination coefficient ρ_i. It is seen that recombination inhibits expansion at higher currents and may even cause further contraction, but the experimental evidence is that recombination plays little part for these conditions.

The contraction in rare gases has been most extensively studied by Kagan and co-workers, with careful simultaneous measurements of the radial variation of electron temperature, electron number density, and

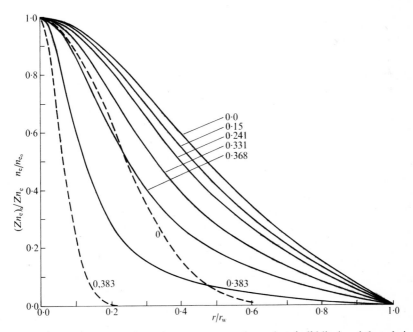

FIG. 5.13. Radial variation of the electron concentration n_e/n_{eo} (solid line) and the relative ionization rate $Zn_e/(Zn_e)_0$ (dashed line) as a function of the parameter $\beta_z = (\rho_i n_e/Z)_0$ which is a measure of the importance of recombination, for the case $C = 1$.

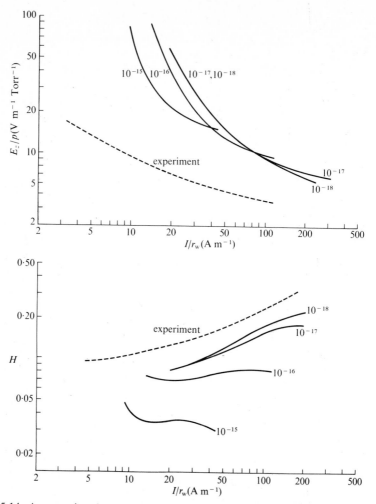

FIG. 5.14. A comparison between theory (solid line) and experiment (dashed line) for (a) the reduced longitudinal field E_z/p and (b) the half-width H as a function of reduced current I/r_w in argon, $pr_w = 500$ mm Torr. The parameter for the solid curves is the effective recombination coefficient $\rho_i(m^3\ s^{-1})$.

light emission intensity as a function of gas pressure and current in terms of the similarity variables pr_w and I/r_w.† As is to be expected the radial variation of electron temperature increases as the current density in the discharge increases, and also as pr_w increases. Fig. 5.15 shows the

---

† The use of I/r_w as a 'similarity' variable under conditions where conventional similarity is no longer applicable has been given some justification by Wojacek (1966). The argument is essentially that it allows for the 'contraction' implied in the model corresponding to Fig. 5.12.

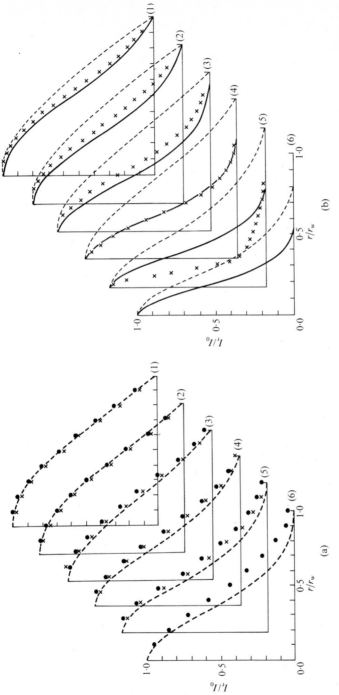

FIG. 5.15. Measurements of the relative radial variation of line intensity I_r/I_0 in neon for (a) $pr_w = 60$ mm Torr, (b) $pr_w = 600$ mm Torr for a sequence of values of I/r_w, 1·6, 0·42, 0·84, 4·2, 8·4, 2·10, 16·7 A m^{-1}, compared with theory for a radially uniform gas temperature (dashed line) and taking into account gas heating (solid line).

measurements of Golubowski, Kagan, and Michel (1970) in neon, indicating that, at $pr_w = 60$ mm Torr, the radial variation of electron temperature is relatively unimportant and that, while the contraction in number density is slight, that in light emission is more clearly seen. At higher currents the subsequent expansion is discernible. On the other hand, at $pr_w = 600$ mm Torr, inclusion of the variation of electron temperature is necessary to obtain agreement between theory and experiment at low current densities, and the relative discontinuity shown in Fig. 5.7 marks a point of departure between experiment and theory.

A further examination of the contraction process has been carried out recently by Smits and Prins (1975). They conclude that only changes in the relative proportions of slow and fast electrons in the distribution function can explain the suddenness of the phenomenon and accordingly set up a model which incorporates a composite distribution function corresponding to Maxwellians at different temperatures T_s and T_f, where $T_s/T_f = R_T$. The parameter R_T is made to depend on the degree of ionization consistent with the expectations of eqn (1.13a) and is discussed in Appendix I (p. 223). In this way a sudden contraction and also a hysteresis between increasing and decreasing current due to gas heating can be predicted theoretically, say by computing the axial field strength E as a function of the discharge current I. On this basis a self-consistent calculation which allowed R_T to be determined *ab initio* should be capable of modelling the phenomenon completely as observed experimentally.

5.9. Contraction in electron-attaching gases

The treatment of the diffuse positive column in electronegative gases given earlier showed that, under suitable conditions (namely, high attachment rates), the central negative ion density can exceed that of the electrons by many times. It was also noted that the radial electron distribution showed a point of inflection. This latter factor is not sufficiently large to account for the sharply contracted columns which are the more usual state for group-VII elements and compounds such as sulphur hexafluoride, in which dissociation may play an important part.

Measurements in such gases are difficult, since the conditions for normal electrostatic probe techniques are not met owing to the presence of negative ions and the discharges themselves are notoriously unstable. Since a significant fraction of the current is carried by the ions, both positive and negative, it is likely that gas heating effects will set in at much lower pressures than for the noble gases and therefore that, at what are normally regarded as low pressures, conditions approximate most closely to those of high pressure ones with a free-standing column in which thermal ionization and conduction are the dominant processes (see e.g. Elenbaas 1951). The most detailed work on contracted columns in these gases appears to be that of Woolsey, Emeleus, Gray, and Coulter (1967), while a collection of papers including this one is available in book form, Emeleus and Woolsey (1970).

5.10. The afterglow positive column

So far we have considered the time-invariant positive column of a steady direct-current discharge. In a number of experiments it has proved convenient to use the afterglow of a discharge. The afterglow is the decaying plasma which exists for time scales typically of the order of milliseconds after the cessation of axial current flow through a steady discharge. This time scale is set by the loss processes operating, which in general will be motion to the walls (either free-flight or collision-dominated) and volume recombination. The decay of free charge density provides a technique for measuring both recombination and diffusion coefficients and has met with considerable success in determining these physical constants for several gases, but care has to be taken when analysing the results to distinguish between the two types of loss mechanism.

The ambipolar diffusion coefficient was introduced in Chapter 2 in connection with the collision-dominated quasi-neutral positive column. In Chapter 4 it was shown that, as the charge-particle density decreases—or more accurately as the ratio λ_D/L of Debye length to discharge dimension increases—the implicit ambipolar assumptions of equal radial velocities no longer holds and the different types of charged particles diffuse independently. This transition was calculated by Allis and Rose (1954), and their results, replotted in terms of λ_D/L, for hydrogen are given in Fig. 5.16.

Sensitive microwave methods of detecting charged-particle densities have allowed the transition to be explored experimentally, and the results of Weaver and Freiberg (1968) at two different pressures in helium are given in Fig. 5.17.

The principal virtue of the afterglow for plasma measurements is that it provides a plasma with a lower noise level and significantly reduced collision rate compared with the active discharge from which it derived. This is due to the fact that the time scale τ_T for decay of electron temperature given approximately by $\tau_T = 3/2\kappa_e \nu_e$ (from eqn (1.14)) is shorter typically by three orders of magnitude than that for decay of number density. The reduced electron temperature implies collision frequencies reduced by an order of magnitude, and also the effects of particle drift inevitably present in an active discharge are avoided.

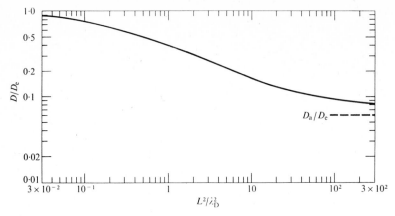

FIG. 5.16. The influence of space charge on particle loss by diffusion as shown by the ratio of the diffusion coefficient D to the electron self-diffusion coefficient D_e as a function of the ratio of the lateral dimension L to the Debye length λ_D for hydrogen gas. For small L/λ_D, $D \to D_e$ and for large L/λ_D, $D \to D_a$.

An appropriate model of the positive column under these conditions can be constructed if one assumes a phase of self-similarity, to the extent that the other discharge parameters remain constant on the time scale that the charged-particle density is varying (i.e. we ignore all time variations except that of density) then we have in the plasma approximation

$$\frac{\partial n}{\partial t} + \frac{\partial}{\partial x}(nv) = -\rho_i n^2,$$

where ρ_i is the recombination coefficient, it being supposed that direct electron–ion recombination is the dominant process.

The equations for electron and ion motion are

$$Mv\frac{dv}{dx} + Mv\nu_i = eE - \frac{1}{n}\frac{d}{dx}(nk_B T_i),$$

$$mv\nu_e = -eE - \frac{1}{n}\frac{d}{dx}(nk_B T_e).$$

In the limit of high collision frequencies there is simplification in that the electric field is given by

$$E = \frac{(M\nu_i T_e - m\nu_e T_i)v}{e(T_e + T_i)} \equiv \frac{(\mu_e T_e - \mu_i T_i)v}{\mu_e\mu_i(T_e + T_i)} \sim \frac{v}{\mu_i},$$

and the number density by

$$-\frac{1}{n}\frac{dn}{dx} = \frac{m\nu_e + M\nu_i}{k_B(T_e + T_i)}v \equiv \frac{e(\mu_e + \mu_i)v}{\mu_e\mu_i k_B(T_e + T_i)} \sim \frac{ev}{\mu_i k_B T_e},$$

and writing this as v/D_a we have

$$\frac{\partial n}{\partial t} - \frac{\partial^2}{\partial x^2}(D_a n) = -\rho_i n^2.$$

If the boundary condition adopted is $n = 0$ at the wall then there will always be a diffusion layer (§ 4.7) adjacent to it, since recombination varies as n^2, whereas the diffusion loss is slower. At high gas pressure and particle densities, the central region will be one of uniform density and in it n will be a constant. Such a model will be valid until the diffusion layer reaches the centre, or until n_0 reaches the value $D_a/\rho_i x_w^2$, thereafter diffusion losses will dominate and be given in plane geometry by

$$n = n_0\left[\left(a_1 e^{-t/\tau}\cos\frac{\pi x}{2x_w} + a_3 e^{-3t/\tau}\cos\frac{3\pi x}{2x_w} + \cdots\right),\right.$$

where $1/\tau = D_a \pi^2/4x_w^2$; this rapidly decays from an initial configuration which defines $a_1, a_3, \ldots$ to a fundamental cosine-like model. The extension to other discharge shapes is readily made, e.g. in cylindrical geometry is given by $r_w^2/D_a(2\cdot405)^2$, and generally one can write $\tau = \Lambda^2/D_a$, where Λ is the diffusion length defined by the particular shape of the region containing the afterglow.

A clear demonstration of the difference in radial distributions between diffusion-controlled and recombination-controlled conditions has been

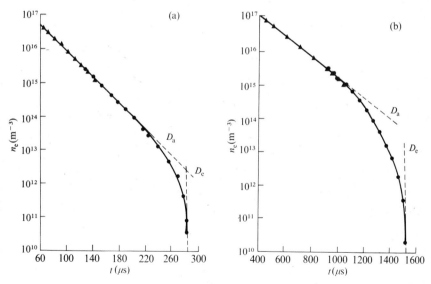

FIG. 5.17. Measured mean electron density $\bar{n}_e$ as a function of time t in a helium afterglow at two pressures (a) 0·4 Torr, (b) 4·0 Torr showing the transition from ambipolar to free diffusion as the density decreases.

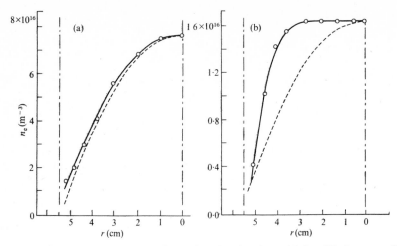

FIG. 5.18. Radial electron-density distributions in afterglows; (a) for diffusion-controlled conditions $p = 0.05$ Torr Kr, $\tau = 7.0$ ms; (b) for recombination-controlled conditions $p = 0.2$ Torr, Kr:$O_2 = 9:1$, $\tau = 1.0$ ms. The dashed curves correspond to the zero-order Bessel function. $T_e = T_g = 300$ K.

given by Smith, Dean, and Adams (1974) using accurate Langmuir-probe techniques to measure electron density. Their results, given in Fig. 5.18, show a zero-order Bessel-function distribution when diffusion dominates and a flat-topped distribution over most of the radius when recombination dominates.

The detailed temporal variations, which for the recombination phase will be given by

$$\frac{1}{n_0(0)} - \frac{1}{n_0(t)} = -\rho_i t$$

and for the diffusion phase by an approximately exponential decay, are difficult to distinguish between in experiments in which only the peak or average density is measured such as the microwave cavity frequency shift method (Biondi and Brown 1949). Criteria which enable the distinction to be made and the accuracy of the coefficients D_a or ρ_i to be estimated have been determined by Gray and Kerr (1962), and this has led to a critical review of the accepted values in the literature prior to that date. The relative importance of recombination and diffusion at the centre is given by the parameter

$$\beta_D = \frac{\rho_i n_0(0)\Lambda^2}{D_a},$$

and for pure recombination in terms of normalized density $N(r) = n(r)/n(0)$ and time $\tau = t D_a/\Lambda^2$ one would have $1/\bar{N} = 1 + \beta_D \tau$, so that

$d(1/\bar{N}) = \beta_D \, d\tau$. Denoting the experimentally determined value of $d(1/\bar{N})/d\tau$ as β_D^*, Biondi and Brown computed, for different geometries and different initial charged-particle distributions, β_D^*/β_D as a function of β_D and also of f, the factor change in $\bar{N}$ for which the plot of $1/\bar{N}$ versus t was linear within an arbitrary 2 per cent. Fig. 5.19 gives β_D^*/β_D as a function of f for cylindrical geometry and shows that an order of magnitude or more of linearity in $1/\bar{N}$ versus t is required to deduce the dominance of recombination.

At low pressures the ion velocity v is determined by inertia, as we have seen, and thus the equations become more complicated. If volume recombination can be ignored some progress can be made, and the momentum equations can be combined to give

$$Mv\frac{\partial v}{\partial x} = -\frac{1}{n}\frac{\partial n}{\partial x} k_B(T_i + T_e) \quad \text{or} \quad Mv^2 = 2k_B(T_i + T_e)\ln\frac{n}{n_0},$$

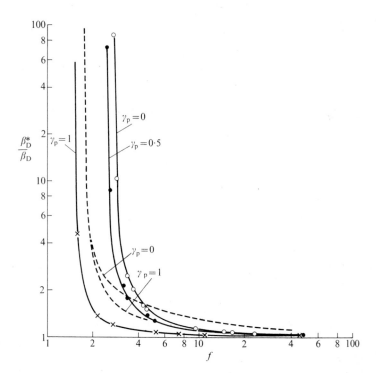

FIG. 5.19. The ratio of the theoretical initial slope β_D^* of the N^{-1} versus time plot to the parameter $\beta_D = \rho_i n_0 \Lambda^2/D_a$ as a function of f the ratio of values of number density over which the plot is linear. Curves are given for different values of γ_p, the ratio of plasma radius to cavity radius and for different initial distributions. Uniform distribution: solid line, Bessel function: dashed line.

whence

$$v\frac{\partial v}{\partial t}+v^2\frac{\partial v}{\partial x}-k_B\frac{T_i+T_e}{M}\frac{\partial v}{\partial x}=0,$$

from which we see that, at the point at which the Bohm criterion is satisfied, velocity is time-invariant. This is a feature which is apparently of general application (Prewett 1975).

Given the ion speed at the column boundary it is possible to estimate the decay time τ_D for the plasma density from the continuity equation

$$\frac{\partial n}{\partial t}+\nabla\cdot n\mathbf{v}=Zn,$$

since with $Z=0$ and using Stokes's theorem

$$-\frac{\partial}{\partial t}\int n\,d\mathbf{S}=\int \nabla\cdot n\mathbf{v}\,d\mathbf{S}=\int n\mathbf{v}\times d\mathbf{S}$$

yields

$$\tau_D=\frac{\bar{n}r_w}{2n_w\{k_B(T_e+T_i)/M\}^{\frac{1}{2}}}.$$

a generalization of the plasma balance equation.

When a magnetic field is applied to the afterglow column the previous considerations will be modified by the presence of the field, as considered in Chapter 3, so that the foregoing discussion applies with the appropriate modification of the ambipolar diffusion coefficient, namely,

$$D_{am}=D_{a0}(\mu_e+\mu_i)\Big/\left\{\mu_e\left(1+\frac{\omega_{ci}^2}{\nu_i^2}\right)+\mu_i\left(1+\frac{\omega_{ce}^2}{\nu_e^2}\right)\right\}.$$

The radial electric field is given by

$$\frac{eE}{k_B T_e}=\frac{1}{n}\frac{dn}{dr}\frac{\mu_e(1+B^2\mu_i^2)-(T_i/T_e)\mu_i(1+B^2\mu_e^2)}{\mu_e(1+B^2\mu_i^2)+\mu_i(1+B^2\mu_e^2)}\equiv\frac{1}{n}\frac{dn}{dr}f,$$

where f ranges from $+1$ to -1 as B increases from 0 to large values ($\gg 1/\mu_i$), and so the electric field is generally reduced by an axial magnetic field. For an afterglow $T_e\sim T_i$ and thus with $\mu_e\gg\mu_i$, $E\sim 0$ when $B^2\sim T_e/\mu_i\mu_e T_i$. Experimental conditions under which it is appropriate to set $E=0$, and therefore $\mathbf{E}\times\mathbf{B}$ drift (i.e. rotation of a cylindrical column) is absent, are made use of in connection with measurements on drift waves as described in Chapter 10.

5.11. Afterglow containing negative ions

In the active positive column the charged-particle distributions were shown to be significantly modified by the presence of negative ions, and

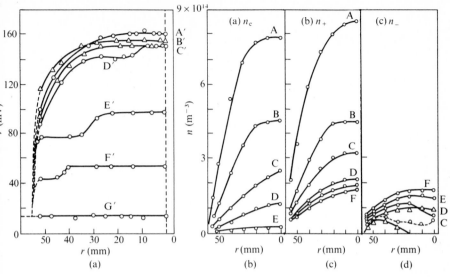

FIG. 5.20. Radial variations of (a) floating potential, (b) electron density, (c) positive ion density, and (d) negative ion density at times A = 0·5, B = 1·0, C = 1·5, D = 2·0, E = 2·4, and F = 3·3 ms A′ = 0·43, B′ = 0·51, C′ = 0·92, D′ = 1·46, E′ = 2·83, F′ = 3·8, and G′ = 4·2 ms in the afterglow of a $Kr:O_2 = 1:1$ discharge showing the development of a different form of variation of the parameters after approximately 1 ms corresponding to the generation of negative ions.

therefore one would expect it also to be true in the case of the afterglow. Attachment and detachment rates are functions of the electron temperature, and therefore the relative concentrations of negative ions and electrons will change with time in the afterglow—thus the properties of discharges containing different proportions of negative ions can conveniently be displayed. Fig. 5.20 shows measurements by Smith, Dean, and Adams (1974) of the space potential and the charged-particle densities at different times in the afterglow of a krypton–oxygen 1 : 1 mixture at 30 mTorr. The negative ion density was deduced by subtraction, and its presence and time scale were confirmed by mass spectrometer. The flattening of the radial potential is similar to that shown in an active discharge (see Fig. 5.6).

5.12. The positive column with an alternating current

It is a fundamental property of the positive column that for a wide range of current densities the discharge parameters are current-independent (see Fig. 4.8, for example). The range as we have seen is determined at low currents by space charge and at high currents by two-stage processes. Within these limits the current can be made to vary with the discharge behaving as a device with constant characteristics. This is true only for sufficiently low frequencies, as can be seen by taking the continuity equation and perturbing it at frequency ω to give

$$i\omega n_1 + n_0\nabla \cdot \mathbf{v}_1 = Zn_1$$

or

$$(\omega + iZ)n_1 = in_0\nabla \cdot \mathbf{v}_1,$$

where the subscript 0 indicates steady state and 1 a quantity varying with frequency ω. The ionization rate Z is seen to define a breakpoint in the frequency response.

Therefore if the current through a steady discharge is modulated at a frequency ω one would expect the number-density modulation to be constant for $\omega \ll Z$ and to drop off at high frequencies with its 3 dB point when $\omega = Z$. The inset to Fig. 5.21 shows measurements by Bryant

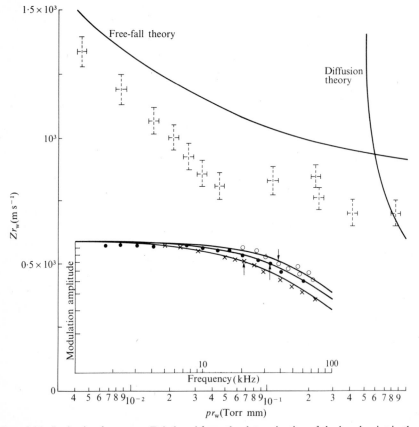

FIG. 5.21. Ionization frequency Z deduced from the determination of the breakpoint in the frequency response of a positive column as indicated in the inset for different pressures. For a range of values of pr_w in mercury, Zr_w is compared with the predictions of the classical free-fall and diffusion theories.

(1966), using the microwave resonance described in § 6.5 to monitor number density, indicating such behaviour. Varying the discharge conditions allows Z to be determined as a function of the parameters varied, and Fig. 5.21 shows Bryant's measurements for mercury as a function of pr_w.

It is beyond the scope of this book to consider the impedance characteristics of the whole discharge, and for details of such measurements the reader is referred to Benson (1965).

6

ELECTRON PLASMA WAVES

6.1. Longitudinal electron waves in infinite uniform plasma

IN CHAPTER 1 the plasma frequency was introduced by supposing that the electrons underwent a uniform displacement. Under these conditions (i.e. of effectively infinite wavelength) oscillation at a frequency $\omega_{pe} = (ne^2/m\varepsilon_0)^{\frac{1}{2}}$ occurred.

If we take the macroscopic or fluid equations of Chapter 1 and consider charged-particle motion, treating the ions as immobile, writing the electron number density† $n_e = n_0 + n_1 \exp\{i(-\omega t + kz)\}$ and the electron velocity and electric field $v_1 \exp\{i(-\omega t + kz)\}$ and $E_1 \exp\{i(-\omega t + kz)\}\mathbf{k}$ respectively, then the terms which vary as $\exp\{i(-\omega t + kz)\}$ in the equations of continuity and motion give

$$-i\omega n_1 + ikn_0 v_1 = 0$$

and

$$-i\omega m v_1 = -eE_1 - k_B T_e\left(\frac{ikn_1}{n_0}\right),$$

while Poisson's equation is $ikE_1 = -en_1/\varepsilon_0$.

These yield the dispersion equation for electron plasma waves in the isothermal fluid approximation

$$\omega^2 = \omega_{pe}^2 + \frac{k_B T_e}{m}k^2. \tag{6.1}$$

It is of considerable interest and importance to consider such waves by taking the Vlasov equation itself, which the distribution function satisfies under collisionless conditions, and perturb it by writing

$$f = f_0(v) + f_1(v)\exp\{i(-\omega t + kz)\},$$

where

$$\int_{-\infty}^{\infty} f_0(v)\,dv = n_0.$$

---

† The convention of writing the propagator for waves $\exp\{i(-\omega t + kz)\}$ is used here. Care is needed when comparing with those authors who use $\exp\{i(\omega t - kz)\}$.

This, together with Poisson's equation, then yields

$$-i\omega f_1 + ivk f_1 - \frac{eE_1}{m}\frac{\partial f_0}{\partial v} = 0,$$

$$+ikE_1 = -\frac{e}{\varepsilon_0}\int_{-\infty}^{\infty} f_1(v)\,dv,$$

giving the dispersion relation

$$k^2 = \omega_{pe}^2 \int_{-\infty}^{\infty} \left\{ \frac{1}{n_0}\frac{\partial f_0}{\partial v} \Big/ \left(-\frac{\omega}{k}+v\right)\right\} dv,$$

which can be rearranged to give the relative permittivity

$$\varepsilon(\omega, k) \equiv 1 - \frac{\omega_{pe}^2}{k^2}\int_{-\infty}^{\infty} \left\{ \frac{\partial f_0}{\partial v} \Big/ n_0\left(-\frac{\omega}{k}+v\right)\right\} dv = 0.$$

The singular integral requires interpretation, and this was first given correctly by Landau (1946), who showed that the path of integration was required to pass below the pole at $v = \omega/k$. This prescription ensures that, for f continuous, $\varepsilon(\omega, k)$ is analytic for complex ω or k, and is justified on the physical grounds of satisfying causality. A detailed discussion is given by Roos (1969).

If f is monotonically decreasing as $|v|$ increases then solutions $\omega + i\gamma$ of $\varepsilon(\omega, k) = 0$ for real k have $\gamma < 0$ and thus are seen to correspond to damped waves. This 'collisionless damping' is commonly referred to as Landau damping.

For $f_0(v)$ a Maxwellian, i.e.

$$f_0 = \frac{n_0}{\pi^{\frac{1}{2}} c_e} \exp\left(-\frac{v^2}{c_e^2}\right),$$

with

$$c_e^2 = \frac{2k_B T_e}{m},$$

the relative permittivity function $\varepsilon(\omega, k)$ can be conveniently expressed in terms of the plasma dispersion function $Z(\omega/k)$ defined by

$$Z(\zeta) = \frac{1}{\sqrt{\pi}}\int_{-\infty}^{\infty} \frac{\exp(-x^2)}{x - \zeta}\,dx,$$

so that introducing Z' the derivative of Z one can write

$$\varepsilon(\omega, k) = 1 - \frac{\omega_{pe}^2}{k^2 c_e^2} Z'\left(\frac{\omega}{kc_e}\right). \tag{6.2}$$

Z and Z' have been computed and tabulated by Fried and Conte (1961).

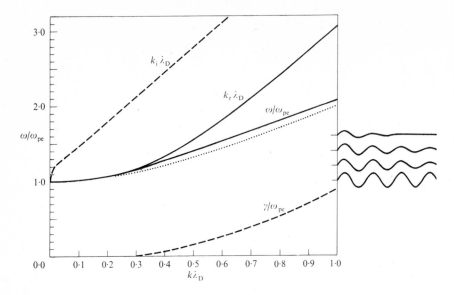

FIG. 6.1. The dispersion relations for electron plasma waves according to kinetic theory, showing complex wavenumber k as a function of real frequency ω, $k = k_r(\omega) + i k_i(\omega)$, and complex ω as a function of real k, $\omega = \omega(k) + i\gamma(k)$. Frequency is normalized to the electron plasma frequency ω_{pe}, and wavenumber to the Debye length λ_D. The real parts are shown full and imaginary dashed. The Bohm and Gross approximation (6.3) is shown dotted.
     Interferograms at ω_{pe}, $1\cdot2\omega_{pe}$, $1\cdot4\omega_{pe}$, and $1\cdot6\omega_{pe}$ show the onset of Landau damping.

Methods of computation more suited to the evaluation of more complicated dispersion relations which arise later have been given by Franklin (1968), Barberio-Corsetti (1970), Ferguson (1971), and Gautschi (1971).
     Z' can be approximated by

$$-2 - i2\sqrt{\pi}\zeta \exp(-\zeta^2)$$

for small ζ, and by

$$\frac{1}{\zeta^2} - i\sigma^2\sqrt{\pi}\zeta \exp(-\zeta^2).$$

where $\sigma = 0$, Im $\zeta > 0$; $\sigma = 1$, Im $\zeta = 0$; $\sigma = 2$, Im $\zeta < 0$; for large ζ.
     Solutions of $\varepsilon = 0$ for complex ω and real k, and complex k and real ω are given in Fig. 6.1† together with the approximation first used by Bohm and Gross (1949), who took the first two terms in the asymptotic

---

† The solutions given correspond to the least damped root of the dispersion relation as this is usually the physically dominant one, but for a contrary case see § 8.8. For a discussion of the higher-order roots of (6.2) see Ecker and Fromling (1974).

expansion of Z' for large ω/k, which yields

$$\omega^2 = \omega_{pe}^2 + \tfrac{3}{2}k^2 c_e^2, \tag{6.3}$$

and apart from the numerical factor is identical with the fluid result. This numerical factor can be recovered in the fluid approximation if a polytropic index γ is introduced so that $\Delta p = \gamma k_B T_e \Delta n$. One then has to consider what is the appropriate value for γ in any given situation (for a discussion of this point see Spitzer (1956)). It can be seen that the damping increases rapidly with k and that one cannot talk of a propagating wave for $k\lambda_D \gtrsim 1/\sqrt{2}$.

The two curves $\omega(k)$ and $k(\omega)$ are given because theoretical workers usually take k to be real and solutions are examined for damping of growth in time, while experimentally it is convenient to excite waves continuously and then measure the spatial dependence of the propagating wave. Fig. 6.1 demonstrates that the difference in the real part of the dispersion relation can be significant.

In this chapter we will be concerned with those experiments in which the propagation of longitudinal electron waves has been studied in plasmas of the type described in Chapters 2 and 3. Such waves have been found propagating both radially across and axially along cylindrical columns and in general it is necessary to take account of the material surrounding the plasma if a complete description of the radiofrequency fields associated with the wave is to be given. It is convenient to formulate the problem in such a way that the plasma is treated as an equivalent dielectric medium, and one can then introduce the relative-permittivity function $\varepsilon(\omega, k)$ given above and use the methods of classical electromagnetism.

If collisions are included in the treatment using a constant-collision-frequency model and we consider the perturbed equation, then the appropriate form is seen to be

$$\frac{\partial f_1}{\partial t} + v\frac{\partial f_1}{\partial z} - \frac{eE_1}{m}\frac{\partial f_0}{\partial v} = -\nu_e f_1 + \nu_e \frac{n_{e1}}{n_0} f_0,$$

since then integrating with respect to the electron velocity gives effectively a statement of the conservation of particles (Bhatnagar, Gross, and Krook 1954). Now taking all quantities to vary as $\exp\{i(-\omega t + kz)\}$ gives

$$(-i\omega + ivk + \nu_e)f_1 - \frac{eE_1}{m}\frac{\partial f_0}{\partial v} = \nu_e \frac{n_{e1}}{n_0} f_0,$$

$$ikE_1 = -\frac{n_{e1}e}{\varepsilon_0},$$

$$n_{e1} = \int f_1 \, dv,$$

and the dispersion relation is found to be given by

$$1 + \frac{i\nu_e}{kc_e} Z\left(\frac{\omega + i\nu_e}{kc_e}\right) - \frac{\omega_{pe}^2}{k^2 c_e^2} Z'\left(\frac{\omega + i\nu_e}{kc_e}\right) = 0$$

or

$$\varepsilon(\omega, k) = 1 - \left\{\frac{\omega_{pe}^2}{k^2 c_e^2} Z'\left(\frac{\omega + i\nu_e}{kc_e}\right)\right\} \Big/ \left\{1 + \frac{i\nu_e}{kc_e} Z\left(\frac{\omega + i\nu_e}{kc_e}\right)\right\}. \tag{6.4}$$

Curves showing the transition from collisional to Landau damping have
been given by Franklin (1968). Axial electron drift v_D is readily included
in (6.4) and modifies the argument

$$\frac{\omega + i\nu_e}{kc_e} \quad \text{to} \quad \frac{\omega - kv_D + i\nu_e}{kc_e}.$$

6.2. Experimental measurements

The propagation of electrostatic waves within the body of an active-
discharge positive column has been studied by Tutter (1968) at 4 mTorr
in argon, and his results for the real part of the dispersion are shown in
Fig. 6.2, together with the Bohm and Gross relation modified by drift and
an approximation to the accurate kinetic curve given in Fig. 6.1. Reasonable agreement is found given the scatter of the measurements. The range
of values of k is approximately $0.2 < k\lambda_D < 0.6$, and, for the purposes of
comparison with finite-geometry treatments in later sections, with the
column radius denoted r_w, $50 < kr_w < 150$.

Coquil, Henry, Le Meur, Castrac, and Treguier (1971), making measurements in a xenon diffusion plasma, were able to confirm both the

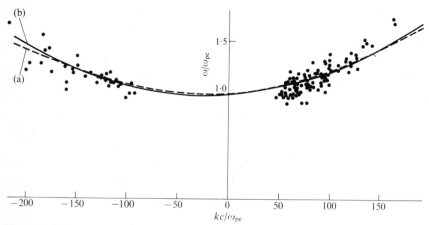

FIG. 6.2. Measurements of the dispersion of electron plasma waves taking into account
electron drift compared with theory: (a) eqn (6.3), (b) an analytical approximation to eqn
(6.2).

imaginary and real parts of the $k(\omega)$ curves of Fig. 6.1 over the range $0\cdot2 < k\lambda_D < 0\cdot5$. Subsequent work (Henry and Treguier 1972) allowed inversion of the process so that the non-Maxwellian nature of the distribution function under certain plasma conditions was deduced from the form of the dispersion curve and confirmed by use of the computational 'multiple-water-bag' technique.

6.3. Axial propagation in a uniform cylindrical column

In general it is true that longitudinal plasma waves have phase velocities small compared with the velocity of light, and this allows the electrostatic approximation to be made, namely that the fluctuating electric fields are given by the gradient of some scalar potential function ϕ satisfying $\nabla\cdot(\varepsilon\nabla\phi)=0$. If the column axis is taken to be in the z-direction then

$$\mathbf{E} = -\nabla[\phi(r,\theta)\exp\{i(k_z z - \omega t)\}],$$

and using the boundary conditions in dielectric media that the tangential components of fields and the radial components of electric displacement are continuous, one can seek to match the conditions inside the plasma to those outside. The most common experimental configurations are for the plasma to be surrounded by a thin (glass) dielectric wall and then by free space, or alternatively, by a metal cylinder or waveguide.

In the first case, assuming the glass wall to be vanishingly thin we have within the plasma $\phi = \phi_1 \exp\{i(k_z z - \omega t)\}$ and outside $\phi = \phi_2 \exp\{i(k_z z - \omega t)\}$, where ϕ is a solution of $(\nabla_T^2 - k_z^2)\phi = 0$; for axisymmetric waves this is in general $AI_0(k_z r) + BK_0(k_z r)$, where I_0 and K_0 are modified Bessel functions.

Since the expression for ϕ_1 must be finite at $r=0$, $B=0$ within the plasma, and since ϕ_2 must tend to zero as $r\to\infty$, $A=0$ in the space outside. Matching E_z and D_r at $r=r_w$ gives

$$AI_0(k_z r_w) = BK_0(k_z r_w)$$
$$\varepsilon AI_0'(k_z r_w) = BK_0'(k_z r_w),$$

or

$$\varepsilon = \frac{I_0(k_z r_w)\,.\,K_0'(k_z r_w)}{I_0'(k_z r_w)\,.\,K_0(k_z r_w)}. \tag{6.5}$$

Asymptotic forms for $I_0(z)$ and $K_0(z)$ are

$$1 + z^2/4 \quad\text{and}\quad -\ln z \quad as\ z\to 0$$

and

$$e^z(2\pi z)^{\frac{1}{2}} \quad\text{and}\quad e^{-z}(\pi/2z)^{\frac{1}{2}} \quad\text{as } z\to\infty,$$

so that using the cold-plasma ($T_e = 0$) approximation we have

$$1 - \frac{\omega_{pe}^2}{\omega^2} \sim \frac{2}{k_z^2 r_w^2} \ln k_z r_w, \qquad\qquad k_z r_w \to 0,$$

$$1 - \frac{\omega_{pe}^2}{\omega^2} \sim -1 \quad \text{or} \quad \omega \to \frac{\omega_{pe}}{\sqrt{2}}, \qquad k_z r_w \to \infty.$$

Thus for small values of k_z we have

$$\frac{\omega}{k_z} \sim \frac{\omega_{pe} r_w}{\sqrt{2}} \ln\left(\frac{1}{k_z r_w}\right),$$

and the electrostatic approximation must break down at values of $k_z r_w \sim \exp(-2c^2/\omega_{pe}^2 r_w^2)$. The shape of the dispersion curve is as in Fig. 6.3, where the dispersion curve for the dipolar mode is also shown. It is given by writing 1 for 0 in the suffixes in eqn (6.5), and can be seen to be a backward wave for a range of k_z, $k_z r_w \lesssim 2$, i.e. the phase velocity ω/k_z and the group velocity $\partial\omega/\partial k_z$ have opposite sign in this range.

Fig. 6.4(a) shows results of a microwave scattering experiment from a positive column in free space (Akao and Ida 1963) showing that when the effect of the glass wall is included good agreement with the theoretical dispersion is obtained. Fig. 6.4(b) shows later results which further clearly

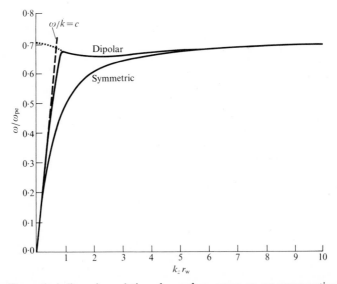

FIG. 6.3. Theoretical dispersion relations for surface waves on an unmagnetized plasma column in free space, showing the symmetric and dipolar modes in the electrostatic approximation. The uncorrected curve at high phase velocities of the order of that of light is shown dotted.

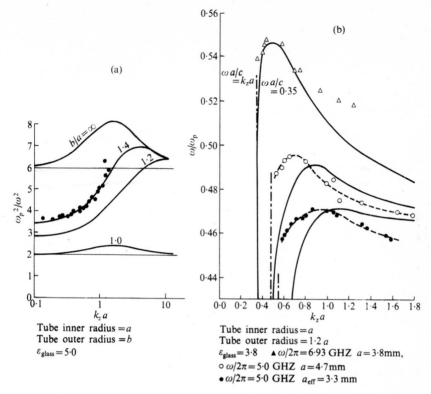

FIG. 6.4. Experimental measurements of surface wave modes (a) by Akao and Ida (1963, 1967) showing comparison with theory taking into account the glass tube, b/a is the ratio of tube outer to inner radius ; (b) by Granastein, Korn, Ojo, and Schlesinger (1967) at long wavelengths when the correction to the electrostatic approximation must be made. $b/a =$ 1·2.

show the need to use the full Maxwell's equations rather than the electrostatic approximation as the phase velocity approaches that of light.

6.4. Axial propagation in a strongly magnetized uniform plasma column

A slow wave solution is not possible in the case where a metal waveguide is filled by an unmagnetized plasma. However, a plasma in a uniform magnetic field is an anisotropic medium and must be described by a tensor relative-permittivity function. Appendix II (p. 224) gives a derivation of the 3×3 matrix form of this Cartesian tensor for a cold plasma which we shall write $\boldsymbol{\varepsilon}$ when appropriate. Thus, when the plasma is in a strong axial magnetic field, such that $\omega_{ce} \gg \omega$, the axial relative permittivity tends to $1 - \omega_{pe}^2/\omega^2$, and the transverse relative permittivity

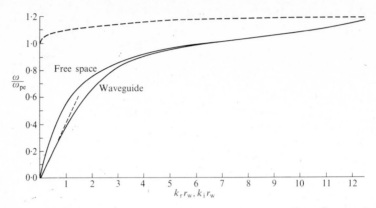

FIG. 6.5. Theoretical dispersion relation $k = k(\omega)$ for waves on a uniform plasma column in a very strong magnetic field for the column in free space and in a metal waveguide. The parameter $\omega_{pe} r_w / c_e = 30$. The linear low-frequency approximation (eqn (6.6a)) for the waveguide case is shown. The spatial damping coefficient $k_i r_w$ is shown dashed and increases rapidly above the plasma frequency.

tends to 1, so that the 'wave' equations becomes

$$\nabla_T^2 \phi - k_z^2 \, \varepsilon \phi = 0, \qquad \varepsilon = 1 - \frac{\omega_{pe}^2}{\omega^2}.$$

For $\varepsilon < 0$ or $\omega < \omega_{pe}$ then there are azimuthally symmetric solutions of the form

$$\phi = J_0(k_z r \sqrt{|\varepsilon|}),$$

with the boundary condition $\partial \phi / \partial z = 0$ at $r = r_w$, giving

$$k_z^2 r_w^2 \left(\frac{\omega_{pe}^2}{\omega^2} - 1 \right) = \beta_{0n}^2, \qquad (6.6)$$

a dispersion equation of the form shown in Fig. 6.5 with a low-frequency phase velocity

$$\frac{\omega}{k_z} = \frac{\omega_{pe} r_w}{\beta_{0n}} \qquad (6.6a)$$

and a cutoff at $\omega = \omega_{pe}$. β_{0n} is the nth zero of the zero-order Bessel function.

The extension to the case where the plasma column is separated from metal walls by a vacuum region is readily treated, and in the vacuum region ϕ is of the form $A I_0(k_z r) + B K_0(k_z r)$. The boundary conditions $E_z = 0$ at $r = r_0$ (r_0 being the outer radius) and E_z and D_r continuous at $r = r_w$ give a more complicated form to the dispersion equation and one dependent on the ratio r_w / r_0.

In the case where $r_0 \to \infty$ it is

$$\frac{\sqrt{|\varepsilon|}\, J_0'(k_z r_w \sqrt{|\varepsilon|})}{J_0(k_z r_w \sqrt{|\varepsilon|})} = \frac{K_0'(k_z r_w)}{K_0(k_z r_w)}, \tag{6.7}$$

and at low frequencies is similar to the unmagnetized case.

Fig. 6.5 gives dispersion relations for finite plasma columns in free space with the magnetic field taken to be infinite. It is under such conditions that the most accurate measurements of wave propagation and damping, demonstrating the existence of Landau damping, have been carried out, but the plasmas used have not been discharges and the reader is referred to Motley (1975) and to the original reports, e.g. Malmberg and Wharton (1964) and Franklin, Hamberger, Lampis, and Smith (1975).

The case where the plasma frequency and electron cyclotron frequency are comparable was considered first both experimentally and theoretically by Trivelpiece and Gould (1959) in the cold-plasma approximation and is dealt with in a later section.

6.5. Radially propagation longitudinal waves in an unmagnetized column

In this section we shall consider the situation in which the perturbation of the plasma column is assumed to be axially uniform. If the electrostatic approximation were appropriate then we would be seeking solutions to $\nabla \cdot (\varepsilon \, \nabla \phi) = 0$, which in the cold-plasma approximation would give, with $\varepsilon = 1 - \omega_{pe}^2/\omega^2$,

$$-\frac{1}{\omega^2}\frac{d\omega_{pe}^2}{dr}\frac{d\phi}{dr} + \left(1 - \frac{\omega_{pe}^2}{\omega^2}\right)\nabla^2\phi = 0.$$

This potential would then have to be matched to that outside the plasma which, again in the electrostatic approximation, would have to satisfy $\nabla^2\phi_0 = 0$. Most experimental situations in which radially propagating waves have been studied have been those in which excitation is by an effectively uniform external field, so that ϕ_0 is of the form $-E_0 x \equiv -E_0 r \cos\theta$. The fields must vary as $\cos\theta$, and, restricting ourselves to the case of a uniform plasma column for the moment, the internal potential must be of the form $Ar\cos\theta$ while the external potential is perturbed by a term of the form $(B/r)\cos\theta$. Matching of tangential electric field strength and radial electric displacement gives

$$A = -2E_0 \Big/ \left(2 - \frac{\omega_{pe}^2}{\omega^2}\right), \qquad B = -E_0 r_w^2 \frac{\omega_{pe}^2}{\omega^2}\Big/ \left(2 - \frac{\omega_{pe}^2}{\omega^2}\right).$$

In this approximation then the fields become infinite for $\omega_{pe}^2 = 2\omega^2$, and so, for instance, in an experiment in which radiofrequencies of

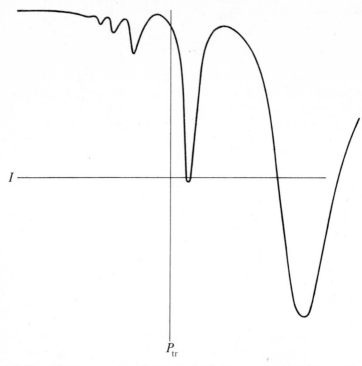

I

P_{tr}

Fig. 6.6. The microwave power P_{tr}, transmitted past a plasma column acting as a capacitative obstacle in a waveguide (perpendicular to the **E** field) showing the pattern of resonances observed by Dattner as the current I through the column was varied so that the plasma frequency exceeded the microwave frequency.

wavelength long compared with column radius propagating perpendicular to the column axis are scattered from a plasma column, one would expect a resonance in the scattering for $\omega_{pe} = \omega\sqrt{2}$ corresponding to the $k = 0$ electrostatic limit in Fig. 6.3. Fig. 6.6 shows the results of such an experiment indicating such a feature. It can be seen also that there are subsidiary resonances in addition to the principal one, and it is possible by varying the dielectric surroundings of the plasma column to show that these are dependent on conditions within the plasma alone (Bryant and Franklin 1963). Extending the treatment above to include the non-uniformity of the column is mathematically tedious but can be shown to give a resonance modified by some mean value of the plasma frequency being $\sqrt{2}$ times the field frequency. Apparently the resonances arise from waves propagating across the plasma column.

The simplest method of treating such waves, yet including their dispersive properties, would be to use the Bohm and Gross dispersion relation.

In which case

$$\varepsilon \approx 1 - \frac{\omega_{pe}^2}{\omega^2} - \frac{3k_B T_e}{m\omega^2} k_r^2,$$

and for a *uniform* plasma the equation $\varepsilon \nabla^2 \phi = 0$ now has to take account of the fact that within the plasma ϕ is oscillatory with wavelength $2\pi/k_r$, so that ϕ satisfies

$$(\nabla^2 + k_r^2) \nabla^2 \phi = 0,$$

where k_r is given by the Bohm and Gross relation. Writing

$$k_r^2 = \frac{(\omega^2 - \omega_{pe}^2)m}{3k_B T_e}$$

and requiring the perturbed electron velocity to be zero at the wall gives resonances for

$$\omega^2 = \omega_{pe}^2 \left(1 + \frac{3\lambda_D^2}{r_w^2} \cdot q_n^2\right),$$

where the q_n are eigenvalues determined by the boundary condition. The sequence of resonances predicted in this way does not agree with experiment for two reasons:

(1) no account has been taken of the radial non-uniformity of the plasma density; and

(2) consequentially in regions where $\omega_{pe} > \omega$ wave propagation cannot occur.

A model overcoming these limitations proposed by Gould, who gave the uniform treatment just described (Gould 1960), and developed by Crawford (1963), uses the WKB (Wentzel–Kramers–Brillouin) approximation to treat wave propagation in the non-uniform plasma. This assumes that the plasma can be treated as locally uniform and that the change of phase $\Delta \phi$ can then be found by integration. Thus

$$\Delta \phi = \int_{r_1}^{r_2} k_r \, dr = \int_{r_1}^{r_2} \left\{ \frac{m(\omega^2 - \omega_{pe}^2)}{3k_B T_e} \right\}^{\frac{1}{2}} dr. \tag{6.8}$$

The range of integration extends from the point r_1 at which $\omega = \omega_{pe}$ to the wall or $r_2 = r_w$. We might expect resonance when this phase change is $m\pi$, m an integer. Taking account of the Stokes phenomenon at r_1 modifies this to $(m + \frac{3}{4})\pi$.

Comparison with experiment shows that the essential features of the subsidiary resonances are adequately explained in that (1) a finite number are predicted, and (2) their separation decreases as the order increases.

The agreement between experiment and theory is surprising in view of the fact that the WKB approximation has been employed in a situation in which the properties of the medium vary so strongly that they change significantly in one wavelength. Better agreement with experiment has been demonstrated recently by How and Blevin (1976) who show that the boundary condition $E_r = 0$ at $r = r_w$ gives resonance for $(m + \frac{1}{4})\pi$, $m = 1, 2, 3 \ldots$.

Using the fluid equations it is possible to set up a model for the problem which takes account of the radial non-uniformity of the steady-state density and leads to an equation of the form

$$\left[\nabla^2 + \frac{\{\omega^2 - \omega_{pe}^2(r)\}m}{3 k_B T_e}\right]\nabla^2\phi =$$

$$= \frac{1}{3n(r)}\left(\nabla n \cdot \nabla(\nabla^2\phi) + \nabla^2 n \, \nabla^2\phi - \frac{(\nabla n)^2}{n}\nabla^2\phi + \frac{\nabla n \cdot \nabla\phi}{\lambda_{De}^2}\right). \quad (6.9)$$

This has been solved numerically by Nickel, Parker, and Gould (1964) subject to boundary conditions ϕ finite at $r = 0$, perturbed electron current vanishing at the wall, and given external field. Fig. 6.7 indicates how good a description can be achieved in this way. Fig. 6.8 gives the perturbed charged-particle distribution corresponding to various-order resonances, showing clearly standing space-charge waves between the point r at which $\omega_{pe} = \omega$ and the wall.

Even more detailed models have been set up, and the reader is referred to Vandenplas (1968) for an extended exposition.

6.6. Damping of radially propagating waves

The existence of resonances in scattering such as those shown in Fig. 6.6 suggests that their height and half-width might be used to determine the magnitude and nature of the damping of the radially propagating electron waves. The possible sources of damping are: (1) collisions, (2) radiation damping, (3) Landau damping, and (4) absorption at the singularity where $\omega = \omega_{pe}$.

In the case of the principal resonance (3) and (4) will not be operative since in a low-pressure active discharge $n_w/n_0 \simeq 0.69$ (see Chapter 2), and the influence of the glass wall is to raise the value of ω_{pe}^2/ω^2 at resonance so that $\omega < \omega_{pe}$ throughout the column. Thus we are left with (1) and (2), and accurate measurements are possible here when the plasma column is made a capacitive obstacle in a waveguide. The reflection coefficient is then determined by the parameter

$$X = \frac{\nu_e}{\omega}\frac{k_g b}{k_0^2 \pi r_w^2},$$

where k_0 is the free-space wavenumber and k_g the wavenumber in the rectangular waveguide of narrow dimension b and r_w is the column

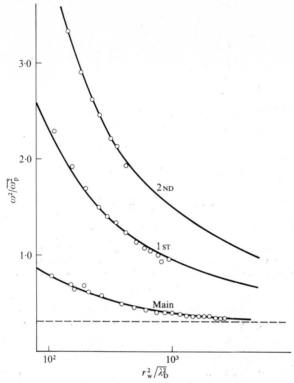

FIG. 6. . Comparison between theory and experiment for the main resonance and the first two subsidiary resonances showing the normalized frequency squared ω^2/ω_{pe}^2 versus the normalized radius squared r_w^2/λ_D^2, $r_w = 3$ mm, $T_e = 3.7$ eV. (Nickel, Parker, and Gould 1964.)

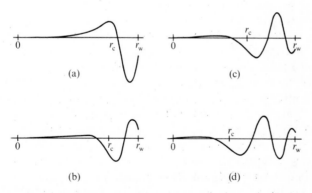

FIG. 6.8. The calculated perturbed electron-density distribution corresponding to the first four subsidiary resonances for the parameter value $r_w^2/\lambda_D^2 = 1600$. r_c marks the point at which $\omega = \omega_{pe}$, r_w indicates the wall.

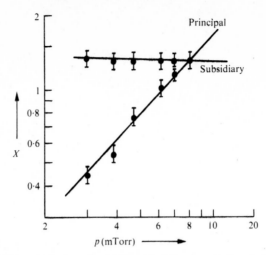

FIG. 6.9. The parameter X, the ratio of plasma damping to radiation damping for a plasma column across a waveguide as a function of gas pressure, showing collisional damping of the principal resonance and 'collisionless' damping for the first subsidiary resonance (mode (a) of Fig. 6.8).

radius. The dependence on b, r_w, k_g, and k_0 was verified by Franklin (1964), and Fig. 6.9 shows the variation of X for the principal or medium resonance as a function of discharge pressure, demonstrating collisional damping. Also shown is the variation of X for the first subsidiary or radial wave resonance, indicating that it is larger and not a function of pressure. The collisionless nature of the damping for the radial waves was confirmed by Prinzler (1969).

Given that the plasma-wave wavelength and hence phase velocity varies significantly in one wavelength in such a strongly radially inhomogeneous plasma it is difficult to see how, on the basis of the resonant-energy-exchange model (O'Neil 1965) of Landau damping, that process can be responsible in this case. Attempts to calculate its magnitude (Jackson and Raether 1966) and experimental observations (Huggins and Raether 1966), however, have suggested otherwise.

A more rigorous approach to the whole problem by Baldwin (1969) and confirmed generally by Baldwin and Ignat (1969), and in greater detail by Ignat (1970), indicates that there is significant absorption in the vicinity of the point where $\omega = \omega_{pe}$. The absorption arises mathematically from a correct treatment of the singularity in the solution of $\nabla^2(\varepsilon\phi) = 0$ through the point where $\varepsilon \to 0$; physically the oscillating field causes the generation of electron plasma waves which then propagate away from the resonant region, and energy is then absorbed elsewhere in the plasma. A comparison between theory and experiment for frequencies near the principal resonance of an afterglow plasma is shown in Fig. 6.10. A later

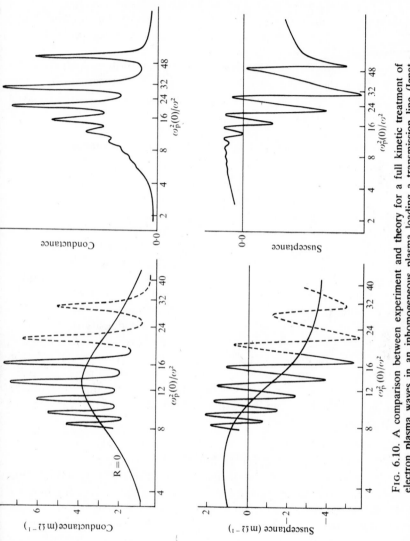

Fig. 6.10. A comparison between experiment and theory for a full kinetic treatment of electron plasma waves in an inhomogeneous plasma loading a transmission line (Ignat 1970). The theoretical calculations were carried out for a zero-order Bessel function radial electron density distribution with $r_w \omega/c_e = 866$ and for $\omega/2\pi = 300$ MHz. The experimental measurements were made after 1 ms in the afterglow of a neon discharge at 0·5 Torr and at $\omega/2\pi = 500$ MHz.

treatment which retains the essential features of that of Baldwin but is one degree lower in the equation involved has been given by Peratt (1973).

This process of resonant absorption is one which can occur in a number of plasma situations and is of considerable current interest in connection with the interaction of intense electromagnetic radiation and inhomogeneous plasmas.

A good summary of work relevant to the situation described in this section and showing the parallels with other physical situations where dissipationless absorption occurs, e.g. Landau damping, has been given by Crawford and Harker (1972).

6.7. General propagation of longitudinal waves in a column

So far discussion has been restricted to the cases in which the wave propagation is (1) purely axial and (2) purely radial. This has been a matter of mathematical convenience, and fortunately it has been possible for experimental configurations to approximate closely to those cases. The microwave scattering experiments of Akao and Ida (1963) showed evidence of the sort of spectrum associated with the radial waves described in the last section, and it is to be expected that a general treatment of waves propagating along a column with a radial variation of number density would confirm this behaviour. Figs 6.11(a) and (b) show the results of computations by Crawford and Tataronis (1965b) for a one-dimensional or slab model with a Gaussian number-density profile giving separately symmetric and antisymmetric modes. It is seen that in addition to the modes of Fig. 6.3, modified by the effect of finite electron temperature, there are modes which in the $k_z = 0$ limit correspond to the resonance for radial waves. Thus all the qualitative features of the scattering experiments can be explained when the radial variation of number density is included.

6.8. Axial propagation in a finite magnetic field

The theoretical work described above has, with the exception of the cold-plasma treatment by Trivelpiece and Gould, all been in the limiting cases where the axial field B_z is zero or infinite. The case of axial propagation where ω_{ce} and ω_{pe} are comparable is given at low phase velocities by $\nabla \cdot (\varepsilon \nabla \phi) = 0$, where using cylindrical coordinates, assuming a uniform column, and taking $\phi = \phi(r)\exp\{i(-\omega t + k_z z)\}$ one finds

$$\left\{\left(1 - \frac{\omega_{pe}^2}{\omega^2 - \omega_{ce}^2}\right)\left(\frac{\partial^2}{\partial r^2} + \frac{1}{r}\frac{\partial}{\partial r}\right) - k_z^2\left(1 - \frac{\omega_{pe}^2}{\omega^2}\right)\right\}\phi = 0$$

having taken account of the difference between relative permittivities

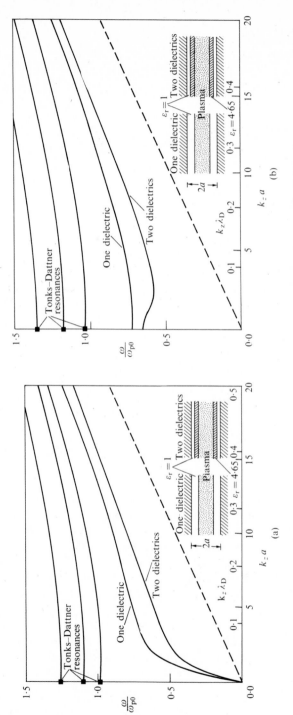

Fig. 6.11. Computed dispersion curves for a plasma slab surrounded by dielectric and between metal plates corresponding to (a) symmetric potentials, (b) antisymmetric potentials showing the radially propagating modes (Tonks–Dattner resonances) as the long-wavelength limit.

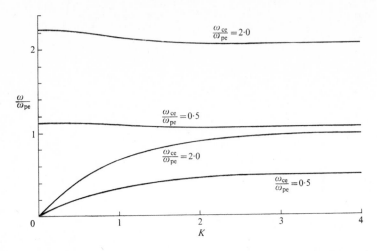

FIG. 6.12. The cold-plasma dispersion relation for waves propagating along the magnetic field for a plasma column surrounded by a metal waveguide and in an axial magnetic field. The axial wavenumber k_z is normalized to the column radius r_w, since $K = k_z r_w / \beta_{01}$, β_{01} being the first zero of the Bessel function J_0. There are two branches with resonant frequencies ω_{ce}, the cyclotron frequency, and ω_{pe}, the electron plasma frequency.

along and transverse to the magnetic field as is shown in Appendix III (p. 229).

For the plasma filling a metal waveguide of radius r_w the dispersion relation is given by

$$\frac{k_z^2 r_w^2}{\beta_{0n}^2} = -\frac{\omega^2(\omega^2 - \omega_{ce}^2 - \omega_{pe}^2)}{(\omega^2 - \omega_{ce}^2)(\omega^2 - \omega_{pe}^2)} \tag{6.10}$$

where β_{0n} is the nth zero of J_0, the zero-order Bessel function of the first kind. The dispersion then depends upon whether ω_{ce} or ω_{pe} is the greater so far as its detail is concerned, but interestingly they enter symmetrically.

Fig. 6.12 shows two cases, one with $\omega_{ce} = 2\omega_{pe}$ and one with $\omega_{ce} = \omega_{pe}/2$ for fixed ω_{pe}. Both the low frequency forward wave and high-frequency backward wave have been studied and their properties verified by Trivelpiece and Gould (1959), as can be seen in Fig. 6.13.

We shall now go on to consider the effect of finite temperature and radial non-uniformity on the radial wave modes which can be excited. It is to be expected that the cold-plasma result for radial propagation being given by $k_z = 0$ in eqn (6.10), i.e. oscillation at the frequency $(\omega_{ce}^2 + \omega_{pe}^2)^{\frac{1}{2}}$, usually called the *upper hybrid* frequency, will remain a feature of such modes.

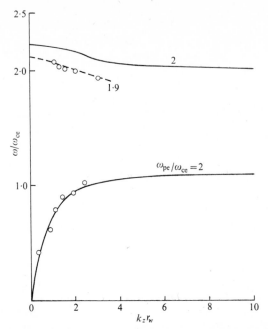

F IG. 6.13. Measurements of electron plasma waves in a magnetized plasma compared with theory for a tube of inner radius r_w and outer radius $1 \cdot 2 r_w$, in a metal tube of radius $2 \cdot 0 r_w$, and K_e (the relative permittivity of the glass) $4 \cdot 6$. Discharge current and magnetic field were such that $\omega_{pe}/\omega_{ce} = 2 \cdot 0$. $\omega_{ce}/2\pi = 250$ MHz, $I = 5$ mA.

6.9. Radial propagation perpendicular to a finite magnetic field—Bernstein waves

In order to consider radially propagating waves, we need to know the relative permittivity for waves propagating transverse to the magnetic field, and this is not a trivial calculation since the appropriate modification to Vlasov's equation now includes a $\mathbf{v} \times \mathbf{B}$ term in the force a particle experiences.

A fluid-approximation treatment can be attempted, modifying the equations in § 6.1 by including the effect of a magnetic field and taking $\mathbf{E} = E_x \mathbf{i} \exp\{i(-\omega t + k_x x)\}$ corresponding to longitudinal waves. The resulting dispersion relation is $\omega^2 = \omega_{pe}^2 + \omega_{ce}^2 + k_x^2(k_B T_e/m)$ and represents an approximation to some parts of the dispersion which results from a kinetic treatment, as we shall see below, but is not generally useful.

A kinetic treatment for waves of this type were first given by Bernstein (1958), and longitudinal waves propagating transverse to a magnetic field are usually known as Bernstein modes. Such a treatment is algebraically complicated and the reader is referred to the original paper or to Stix (1962) for details. The important new physical result is the possibility of propagation near to and slightly above the harmonics of the cyclotron frequency. Introducing the parameter

$$\lambda = \frac{k_x^2}{\omega_{ce}^2} \frac{k_B T_e}{m},$$

the square of the wavenumber times the electron mean Larmor radius, the dispersion relation can be written in alternative forms

$$\varepsilon(\omega, k_x) = 1 + 2\omega_{pe}^2 \frac{\exp(-\lambda)}{\lambda} \sum_1^\infty \frac{n^2 I_n(\lambda)}{(n^2 \omega_{ce}^2 - \omega^2)} = 0, \qquad (6.11a)$$

or

$$\varepsilon(\omega, k_x) = 1 + \frac{\omega_{pe}^2}{\omega_{ce}^2}\left\{1 - \frac{I_0(\lambda)\exp(-\lambda)}{\lambda}\right\} + \frac{2\omega_{pe}^2}{\omega_{ce}^2}\sum_1^\infty I_n(\lambda)\frac{\omega^2}{(n^2\omega_{ce}^2 - \omega^2)}\frac{\exp(-\lambda)}{\lambda} = 0$$

$$(6.11b)$$

The second series converges more rapidly than the first and is more useful in approximating and in numerical work. The process can be extended further, as has been demonstrated by Diament (1967). Dispersion curves for these modes have been calculated by several workers: those given by Crawford and Tataronis (1965a) are reproduced in Fig. 6.14, where the general feature of propagation near the cyclotron harmonics is seen. The fluid approximation has a meaning only for $\omega_{pe}^2/\omega_{ce}^2 > 1$ and $\omega/k < c_e$ on the branch for which, as $k_x \to 0$, $\omega^2 \to \omega_{pe}^2 + \omega_{ce}^2$.

Since in the collisionless case there can be no time-averaged motion of the electrons across the magnetic field lines, there are no particles with directed velocities in resonance with the wave phase velocity, and so Landau damping of exactly perpendicularly propagating Bernstein waves cannot occur.

Propagation of Bernstein waves at an oblique angle to the magnetic field gives rise to an even more complicated dispersion relation than that quoted above for perpendicular propagation. There is, in general, the need to take into account the fact that the electron temperature parallel to the magnetic field $T_{e\parallel}$ and that perpendicular to it $T_{e\perp}$ may not be equal because of the decoupling effect of the field. Detailed results have been given by Crawford (1967), which indicated that for deviations of a few degrees there is strong damping of all branches except for the portions of the curves given by $\omega/k_\perp > c_{e\perp}$ and that the branch passing through the upper hybrid frequency, i.e. that given asymptotically by fluid theory, is the least heavily damped. The requirement for low cyclotron damping is that the phase velocity corresponding to the Doppler-shifted frequency along the field lines should be large compared with the corresponding thermal velocity, i.e.

$$\frac{\omega - n\omega_{ce}}{k_\parallel} \gg c_{e\parallel}.$$

This condition can be used to give the angular requirement

$$\cot \theta = \frac{k_\parallel}{k_\perp} \ll \frac{\omega - n\omega_{ce}}{\omega_{ce}} \cdot \frac{\omega_{ce}}{k_\perp c_{e\perp}} \cdot \frac{c_{e\perp}}{c_{e\parallel}},$$

showing for $k_\perp > \omega_{ce}/c_{e\perp}$ when $\omega \to n\omega_{ce}$ that the angle $\pi/2 - \theta$ is restricted to very small values.

The properties of Bernstein modes were first investigated in situations where it is necessary to include the spatial variation of the plasma frequency, i.e. the electron density, and will be discussed in the next section. However, results are available for a situation which can be compared with Fig. 6.14. These were obtained by Leuterer (1969) in the

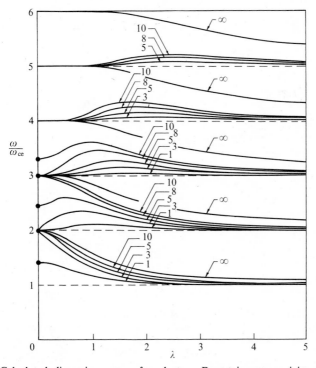

FIG. 6.14. Calculated dispersion curves for electron Bernstein waves giving frequency normalized to the electron cyclotron frequency ω_{ce} as a function of wavenumber normalized to the electron mean Larmor radius for different values of the parameter $((\omega_{pe}/\omega_{ce})^2$. The upper hybrid frequency $\omega_{UH}^2 = \omega_{pe}^2 + \omega_{ce}^2$ is indicated by a spot.

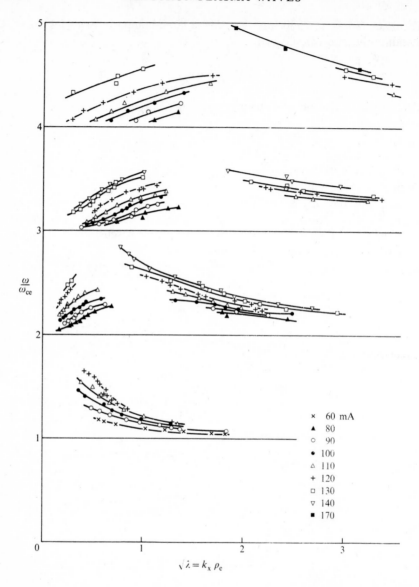

FIG. 6.15. Experimental measurements of the electron Bernstein waves in a low-pressure helium discharge with the plasma frequency varied by varying the discharge current, showing both forward and backward waves and propagation near harmonics of the electron cyclotron frequency up to the fourth.

positive column of a helium discharge at 10 mTorr for a range of current densities, and Fig. 6.15 shows all the principal features of the theoretical curves, in particular the simultaneous existence of both forward and backward waves. The antennae and receiver were near the axis of the discharge and moved over a small range so that it was appropriate to consider the plasma uniform.

6.10. Bernstein waves in a non-uniform plasma

We have seen in § 6.5 that it was possible to analyse the radial propagation of electron plasma waves in a positive column by a suitable modification of the theory for a uniform plasma. 'Standing waves' were then found effectively between the walls and the points at which $\omega = \omega_{pe}$ or $k \to 0$.

Consider the situation for Bernstein waves in a positive column restricting attention to the frequency interval between ω_{ce} and $2\omega_{ce}$. Fig. 6.14 shows that for a uniform plasma in this interval all waves are backward ($\partial\omega/\partial k < 0$) and for fixed ω, with $\omega_{ce} < \omega < 2\omega_{ce}$, as ω_{pe} decreases the wavenumber decreases, becoming zero for $\omega_{pe} = (\omega^2 - \omega_{ce}^2)^{\frac{1}{2}}$. Converted into number density terms in a non-uniform plasma, this means that wave propagation is possible (provided a WKB analysis is valid) from the centre out to the radius r^* at which $n_e(r) = (m\varepsilon_0/e^2)(\omega^2 - \omega_{ce}^2)$. This then is the inverse of the situation for electron plasma waves but would be expected otherwise to follow the same sort of analysis.

In order to make quantitative predictions, and expression is needed for the 'relative permittivity' in the frequency range $\omega_{ce} - 2\omega_{ce}$. This is provided with sufficient accuracy by taking the expression (6.11b) and expanding to first order in λ, using terms up to and including $n = 2$ to give, after some manipulation,

$$\varepsilon(\omega, k) = 1 + \frac{\omega_{pe}^2}{\omega_{ce}^2 - \omega^2} - \left(\frac{3k_B T_e k_r^2}{m}\right) \frac{\omega_{pe}^2}{(4\omega_{ce}^2 - \omega^2)(\omega_{ce}^2 - \omega^2)} \quad (6.12)$$

or

$$\varepsilon = 0 \quad \text{for} \quad k_r^2 = \frac{(\omega_{ce}^2 + \omega_{pe}^2 - \omega^2)(4\omega_{ce}^2 - \omega^2)}{\omega_{pe}^2}\left(\frac{m}{3k_B T_e}\right),$$

and thus the appropriate WKB equation for the wave potential ϕ would be

$$\frac{d^2\phi}{dr^2} + \left(\frac{m}{3k_B T_e}\right) \frac{(4\omega_{ce}^2 - \omega^2)(\omega_{ce}^2 + \omega_{pe}^2 - \omega^2)}{\omega_{pe}^2} \phi = 0.$$

Buchsbaum and Hasegawa (1964) derived this equation by a different but equivalent method and then proceeded to use it in the case where ω_{pe}^2

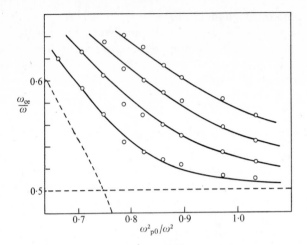

F$_{\mathrm{IG}}$. 6.16. Resonant frequencies in the microwave absorption spectrum for a plasma column compared with theory for standing electron Bernstein waves in the frequency range $\omega_{ce} < \omega < 2\omega_{ce}$ as a function of the normalized peak plasma frequency squared ω_{p0}^2/ω^2. The limiting values corresponding to $\omega = 2\omega_{ce}$ and $\omega^2 = \omega_{pe}^2 + \omega_{ce}^2$ are shown dashed.

was taken to be of the form

$$\omega_{pe}^2 = \omega_{pe0}^2 \Big/ \left(1 + \frac{\gamma_r r^2}{r_w^2}\right),$$

where γ_r is a constant to model a discharge of diameter $2r_w$. With this form ϕ can be found in terms of the Weber parabolic cylinder functions, and imposing boundary conditions gives resonances analogous to those for radially propagating waves in the absence of a magnetic field.

Fig. 6.16 shows a comparison of the results of such an analysis with experimental results obtained by observing peaks in the microwave absorption spectrum of a magnetized positive column. Also shown are the limiting values $\omega = 2\omega_{ce}$ and $\omega^2 = \omega_{ce}^2 + \omega_{pe0}^2$, which together with $\omega = \omega_{ce}$ define the limits of the region within which resonance can recur. Corresponding measurements in transmission have been made by Schmitt, Meltz, and Freyheit (1965) and in emission by Tanaka, Kubo, and Mitani (1965).

The theoretical treatment can be further extended to include the effects of the radial ambipolar field and the method of excitation. Estimates of the contribution of these effects have been given by Pearson (1966) and Frisch and Pearson (1966).

The fact that there is no Landau damping of Bernstein modes means that the observed spectrum is determined by collisional and radiation damping and therefore under appropriate experimental conditions can be

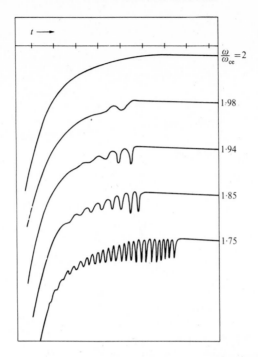

$t \longrightarrow$

$\dfrac{\omega}{\omega_{ce}} = 2$

$1 \cdot 98$

$1 \cdot 94$

$1 \cdot 85$

$1 \cdot 75$

FIG. 6.17. The reflection spectrum from the afterglow of a low-pressure neon discharge showing as a function of time t the sharp resonances associated with standing Bernstein waves and indicating how sensitive the spectrum is to the magnetic field. Experimental conditions: neon 20 mTorr, $\omega/2\pi = 400$ MHz, $r_w = 15$ mm.

made strikingly clear. Fig. 6.17 gives the results of Schmitt, Meltz, and Freyheit, demonstrating this point in an afterglow at 20 mTorr in neon.

7

ION WAVES

7.1. The fluid approximation

IF THE ions in a plasma are not regarded as immobile a separate low-frequency branch of the dispersion relation is found to exist. This is most readily seen by using the fluid equations of continuity and momentum for electrons and ions in the quasi-neutral approximation and ignoring electron inertia and particle generation processes

$$\nabla \cdot (n\mathbf{v}) = 0, \qquad 0 = -e\mathbf{E} - \nabla(nk_B T_e),$$

and

$$M(d\mathbf{v}/dt) = e\mathbf{E}.$$

Setting the number density $n = n_0 + n_1 \exp\{-i(\omega t - kz)\}$ the electric field $\mathbf{E} = E_1 \mathbf{k} \exp\{-i(\omega t - kz)\}$ and the velocity $\mathbf{v} = v_1 \mathbf{k} \exp\{-i(\omega t - kz)\}$ the equations in the perturbed quantities become

$$-i\omega n_1 + ikn_0 v_1 = 0,$$

$$-eE_1 - \frac{ikn_1}{n_0} k_B T_e = 0,$$

$$-iM\omega v_1 - eE_1 = 0.$$

The determinant of these equations gives the dispersion relation

$$\omega^2 = k^2 \frac{k_B T_e}{M} \equiv k^2 c_s^2. \tag{7.1}$$

If quasi-neutrality is not assumed, i.e. $n_e \neq n_i$, the modified equations yield a dispersion relation

$$\omega^2 = k^2 c_s^2 / (1 + k^2 \lambda_D^2); \tag{7.2}$$

and if additionally $T_i \neq 0$ then one finds

$$\frac{\omega^2}{k^2} = \frac{c_s'^2}{1 + k^2 \lambda_D^2} + \frac{k^2 \lambda_D^2}{1 + k^2 \lambda_D^2} \frac{k_B T_i}{M}, \tag{7.3}$$

where

$$c_s'^2 = \frac{k_B(T_e + T_i)}{M}.$$

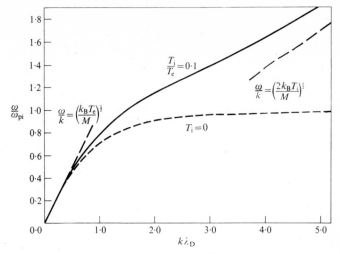

FIG. 7.1. The dispersion relation for ion waves in the fluid approximation in terms of the frequency ω normalized to the ion plasma frequency ω_{pi} and the wavenumber k normalized to the Debye length λ_D. Curves are given for $T_i/T_e = 0, 0.1$; T_i, T_e are the ion and electron temperatures and the asymptotic values of the phase velocity $(k_B T_e/M)^{\frac{1}{2}}$ and $(2k_B T_i/M)^{\frac{1}{2}}$ are shown.

This relation together with the approximations (7.1) and (7.2) is given in Fig. 7.1 for the case $T_i/T_e = 0.1$ and is the fluid approximation to the dispersion relation for ion waves. These waves are also called *ion acoustic* or *electro-acoustic* because below the ion plasma frequency they are characterized by the *electron* temperature and the *ion* mass.

In order to examine the nature of these waves consider the electric field

$$E_1 = -ik \frac{n_{1e}}{n_0} \frac{k_B T_e}{e}.$$

This has an associated potential ϕ_1, where

$$e\phi_1 = k_B T_e \frac{n_{1e}}{n_0}, \tag{7.4}$$

and it can be seen that the wave is essentially a charged-particle density fluctuation since for $k\lambda_D \ll 1$, $n_{1e} \sim n_{1i}$ and thus the electrons may be regarded as following the ion fluctuations. For large values of k the electric fields and the electron perturbation became small and the wave is an ordinary ion sound wave.

It is also of interest to compare the nature of electron and ion waves. For electron waves it is readily seen, using Poisson's equation, that

$$\frac{e\phi_1}{k_B T_e} \sim \frac{n_{1e} e^2}{\varepsilon_0 k^2 k_B T_e} \sim \frac{n_{1e}/n_0}{k^2 \lambda_D^2}, \tag{7.5}$$

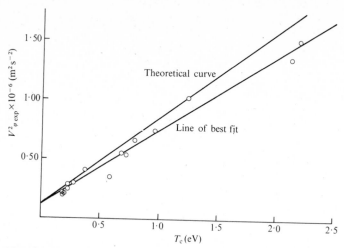

FIG. 7.2. Experimental measurements of the ion wave phase velocity v_ϕ as a function of the electron temperature in xenon ($p \sim 1$ mTorr) compared with the theoretical value. The finite intercept at $T_e = 0$ indicates that the ion pressure and hence temperature influences propagation and can be estimated from it.

and thus the potential for a given density perturbation is larger than that for an ion wave by a factor $(k\lambda_D)^{-2}$ and is thus properly regarded in physical terms as a space-charge wave.

There have been extensive experimental studies of ion waves over the past decade, and the most detailed and comprehensive measurements are those of Alexeff, Jones, and Montgomery (1968). Fig. 7.2 shows how, by measuring the phase-velocity dependence on electron temperature for waves whose wavelength was long compared with the Debye length, they were able to estimate the *ion* temperature in a low-pressure diffusion plasma.

7.2. Kinetic theory of ion waves

Ion waves can be treated as with electron waves by using the Vlasov equation and including ion motion. One then finds an expression for the relative permittivity that is a simple modification of (6.2), by adding a susceptibility term for ions analogous to that for electrons

$$\varepsilon(\omega, k) = 1 - \frac{\omega_{pe}^2}{k^2 c_e^2} Z'\left(\frac{\omega}{kc_e}\right) - \frac{\omega_{pi}^2}{k^2 c_i^2} Z'\left(\frac{\omega}{kc_i}\right), \tag{7.6}$$

c_i being defined by $\frac{1}{2}Mc_i^2 = k_B T_i$. If the phase velocity is comparable to the ion thermal velocity one would expect the phenomenon of Landau damping again to occur, the resonant particles in this case being ions.

For waves with wavelengths long compared to the Debye length the second term is large compared with the first in (7.6) and the dispersion is given by

$$Z'\left(\frac{\omega}{kc_i}\right) \approx \frac{2T_i}{T_e}.$$

The form of the dispersion curve thus depends on the temperature ratio T_i/T_e and for small values can be approximated by using $Z'(x) \sim 1/x^2$ for large x to give consistently

$$\frac{\omega^2}{k^2} \sim \frac{k_B T_e}{M_i}.$$

This, as already noted, is valid only for $k < 1/\lambda_D$ or equivalent $\omega < \omega_{pi}$. As k approaches $1/\lambda_D$ the appropriate form becomes

$$Z'\left(\frac{\omega}{kc_i}\right) = \frac{2T_i}{T_e}(1 + k^2\lambda_D^2),$$

and as the right-hand side increases in magnitude the real part of ω/kc_i decreases towards unity and its imaginary part increases significantly, physically due to Landau damping.

Dispersion curves for the cases $T_i/T_e = 1\cdot0$, $0\cdot1$, $0\cdot01$ are given in Fig. 7.3.

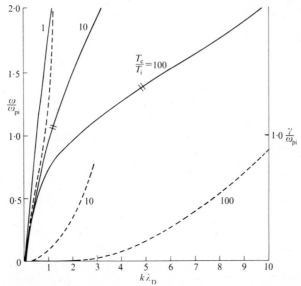

FIG. 7.3. The theoretical dispersion and damping of ion waves given by kinetic theory for values of the temperature ratio T_e/T_i, 100, 10, and 1. The bars indicate frequency limits beyond which the wave is heavily damped ($\gamma/\omega_{pi} > 0\cdot1$), and this is true of all frequencies for $T_i = T_e$.

It can readily be seen from the values of the damping coefficient γ that the damping becomes heavy for $\omega > \omega_{pi}$ or $k\lambda_D > 1$, corresponding to the similar situation for electron waves. Also it is seen that for $T_e = T_i$ it is not meaningful to speak of a wave at all since γ/ω has a minimum value as $\omega \to 0$ of 0.6. This has provoked much discussion of observations on ion oscillations in contact-ionization plasmas, for the resolution of which the reader is referred to Estabrook and Alexeff (1972) and Buzzi (1974).

The effects of collisions and drift can be introduced into the expressions which yield the relative permittivity in the same way as with electron waves (§ 6.1 leading to eqn (6.4)) and result in

$$\varepsilon(\omega, k, v_{De}, v_{Di}, \nu_e, \nu_i) \equiv 1 - \frac{\omega_{pe}^2}{k^2 c_e^2}\frac{Z'\{(\omega - kv_{De} + i\nu_e)/kc_e\}}{1 + (i\nu_e/kc_e)Z\{(\omega - kv_{De} + i\nu_e)/kc_e\}} - \frac{\omega_{pi}^2}{k^2 c_i^2}\frac{Z'\{(\omega - kv_{Di} + i\nu_i)/kc_i\}}{1 + (i\nu_i/kc_i)Z\{(\omega - kv_{Di} + i\nu_i)/kc_i\}} \quad (7.7)$$

It is to be expected that the effect of drift will modify the phase velocity and that for small collisional frequencies there will be an effective addition of collisional and Landau damping. Expressions for the phase velocity and damping for $k\lambda_D \ll 1$ using the approximations given in Chapter 6 for Z and Z', after some manipulation, are found to be

$$\omega/k = v_{Di} \pm c_s,$$

$$\frac{\gamma}{\omega_{pi}} = \frac{\nu_i}{\omega_{pi}} + \sqrt{\frac{\pi}{8}}\, k\lambda_D\left\{\left(\frac{T_e}{T_i}\right)^{\frac{3}{2}}\exp\left(-\frac{T_e}{2T_i}\right) + \left(\frac{m}{M}\right)^{\frac{1}{2}}\left(1 \pm \frac{v_{Di} - v_{De}}{c_s}\right)\right\}. \quad (7.8)$$

7.3. Measurements showing kinetic effects

Measurements of ion wave propagation in a low-pressure active discharge (~ 1 mTorr in mercury) have been made by Sato, Sasaki, Aoki, and Hatta (1967). The ion drift velocity $v_{Di} \sim 5 \times 10^3$ cm s^{-1} was small compared with the ion wave phase velocity $c_s \sim 7 \times 10^4$ cm s^{-1}, and it was observed that waves propagated in opposite directions with different damping as shown in Fig. 7.4. From an analysis of the variation of damping rate with frequency it was concluded that both collisional and Landau damping contributed to the observed spatial decrement. The combined effect of the two processes has been shown most clearly by Alexeff, Jones, and Montgomery (1968). Fig. 7.5 reproduces their results in which the electron temperature was varied while all other plasma parameters were held constant.

The effects of larger drift velocities will be discussed later.

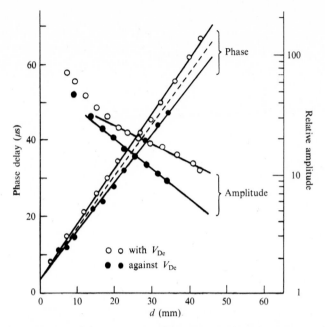

FIG. 7.4. Measurements of the spatial damping and phase delay for ion waves in a drifting plasma at fixed frequency. The dashed line represents the ion acoustic speed c_s corresponding to the measured electron temperature, so that the ion drift v_{Di} is small compared to c_s. The damping is seen to vary with the direction of propagation.

7.4. Finite plasma and other effects

In order to include the real effects of number-density variation and particle generation, let us return to the fluid equations in the quasi-neutral approximation. The picture becomes a little more complicated but is not difficult to anticipate in physical terms.

So far as generation is concerned, the continuity equation becomes

$$-i\omega n_1 + ikv_1 n_0 = Zn_1.$$

Collisions and generation give momentum equations

$$-i\omega v_1 M + Mv_1(\nu_i + Z) = eE_1 ikk_B T_i n_1/n_0,$$
$$-i\omega v_1 m + mv_1(\nu_e + Z) = -eE_1 ikk_B T_e n_1/n_0,$$

to yield

$$(\omega - iZ)\left(\omega + iZ + i\frac{\nu_i M + \nu_e m}{M + m}\right) = k^2 \frac{k_B(T_i + T_e)}{M + m},$$

which in the absence of collisions or when $Z \gg \nu_i$, $(m/M)\,\nu_e$ implies a cut-off $\omega \to 0$ at $k = Z/c_s$ or suggests, using the plasma balance eqn (2.4),

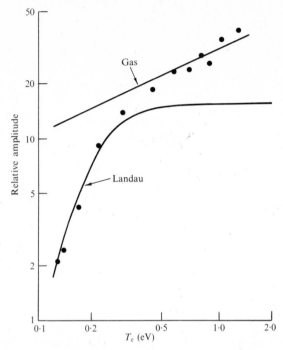

FIG. 7.5. Measurements of ion wave damping as a function of electron temperature by Alexeff, Jones, and Montgomery (1968), showing that a transition can be observed between damping by collisions with the gas atoms and Landau damping. The gas was xenon at 1 mTorr, and the ion temperature $T_i \sim 0 \cdot 05$ eV.

that waves with wavelength less than the radius cannot propagate. This is due to deficiencies in the model in that it has been assumed that the particle temperatures are not subject to perturbation by the wave. Since Z is strongly dependent on T_e, however, it is to be expected that any variation in T_e will be effectively amplified because of this dependence, and this point is pursued in Chapter 8.

Before considering the effects of temperature and radial variations in the steady-state plasma of a positive column it is useful to consider propagation of ion waves on a finite uniform plasma column similarly to the way in which we considered electron waves. The most convenient case to consider is that of a uniform plasma filling a conducting tube, when the dispersion relation can readily be shown, as in § 6.4, to be given by

$$J_m(i\sqrt{\varepsilon}kr_w) = 0,$$

where the fields are assumed to be of the form

$$\exp\{i(m\theta - \omega t + kz)\}$$

and thus the dispersion relation is $\varepsilon + (\beta_{mn}^2/k^2 r_w^2) = 0$, where β_{mn} is the nth zero of the mth-order Bessel function. For the case of $T_e > T_i$, ε can be approximated by

$$1 + \frac{1}{k^2 \lambda_D^2} - \frac{\omega_{pi}^2}{\omega^2}$$

for $\omega < \omega_{pi}$, and thus for small k one finds

$$\frac{\omega^2}{k^2} \approx \frac{k_B T_e}{M} \frac{1}{1 + (\beta_{mn}^2 \lambda_D^2 / r_w^2)}.$$

The phase velocities of higher-order modes are lower and consequently are more heavily damped by Landau damping on the ions, but for $r_w \gg \lambda_D$ the dispersion relations are not significantly different from that for a uniform plasma. This contrasts strongly with the case of electron waves on finite columns where, as we have already seen in Figs 6.3, 6.5, and 6.13, the low-frequency portions of the dispersion curves are significantly modified. Physically this lack of modification in the ion-wave case arises from the fact, already commented on, of the small electric fields associated with the wave. At frequencies above the ion plasma frequency all the modes coalesce into the ion sound wave. The conclusions stated are generally true when $T_e \sim T_i$.

7.5. Ion waves in positive columns

Early experiments to demonstrate the existence of ion waves and measure their propagation were carried out in the positive columns of low-pressure discharges (see e.g. Little (1961), whose results are shown in Fig. 7.6). While the general feature of an approximately linear relation between frequency and wavenumber was confirmed, departures at long wavelengths demanded that the theory should be based on a more complete description of the actual plasma. An appropriate approach then would be to take the models of Chapter 2 and examine low-frequency perturbations of the steady state.

The first step in this direction was made by Woods (1965), who set up the low-pressure ($\nu_e = \nu_i = 0$) model and examined both radial and azimuthal modes of the column. Fig. 7.7 shows that the lowest-order radial and axisymmetric mode $(0, 0)$ is little modified while the experimental points were better approximated by the first-order radial mode $(0, 1)$. This was difficult to understand since that mode was predicted to be more heavily damped than the fundamental. Davies (1966) extended the work by including finite ν_i, and since the equations contain several non-linear terms, the only method of solution available is numerical. Ewald, Crawford, and Self (1969) further took into account the effect of

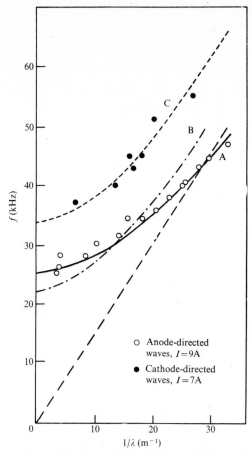

FIG. 7.6. Measurements by Little (1961) of ion waves in positive columns, radius 25 mm, with the waves excited externally and detected in the light output showing a phase velocity close to the ion acoustic speed but indicating that drift and finite radius influence the propagation. Curve A is theory with $r_w = 25$ mm, $T_e = 5 \cdot 2 \times 10^4$ K; while curve B has $r_w = 20$ mm, $T_e = 4 \times 10^4$ K and drift is included; curve C has $r_w = 19$ mm, $T_e = 3 \cdot 7 \times 10^4$ K.

an axial magnetic field and the axial drift which were present in the experiments. They approximated the complete solution by taking the first two terms of a Fourier–Bessel expansion in order to make the problem manageable and concluded that the real part of the dispersion was significantly modified by drift (see Fig. 7.8) and that the damping was likely to be properly accounted for only when the kinetic effect of drift on damping were included in the treatment. This latter is taken up in § 7.7 for a uniform but collisional plasma. Barrett (1966) attempted to estimate the ion drift velocity at low pressures, recognizing, as has been pointed

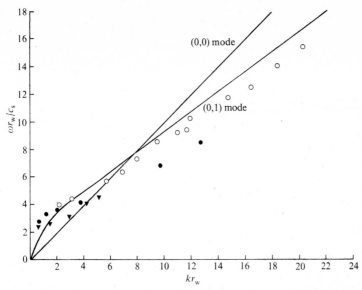

FIG. 7.7. A comparison with theory (Woods 1966) for a low-pressure finite plasma of experimental measurements (Barrett and Little 1965) at low pressure in mercury. Frequency is normalized by r_w/c_s and wavenumber by the plasma radius r_w.

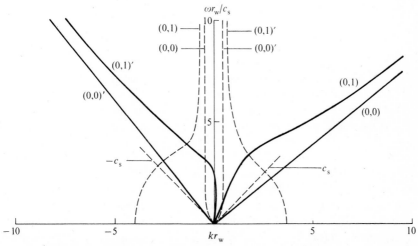

FIG. 7.8. Calculations for the symmetric ion wave mode on a low-pressure mercury column (Ewald, Crawford, and Self 1969) with finite ion drift velocity, indicating that under typical conditions ($\nu_i/Z = \frac{1}{9}$, $\mu_e/\mu_i = 500$, $eE_z r_w/k_B T_e = 0.2$, $\omega_{ce}/(\nu_e + Z) = 100$) the lowest-order mode is the least damped and that drift might explain the consequent discrepancy in Fig. 7.7.

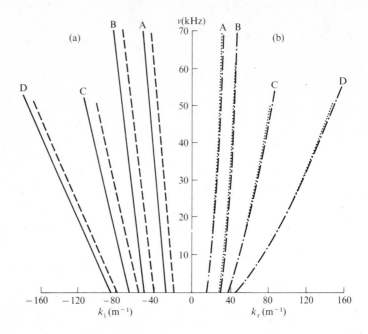

FIG. 7.9. Calculations of the dispersion and damping under conditions appropriate to argon taking into account finite radial variations of all quantities; (a) demonstrating by comparison with corresponding results for a uniform column (b) that such factors have only a small influence on the dispersion. Parameters for A, B, C, D, $pr_w = 0.05$, 0.5, 5, and 50 mm Torr respectively.

out in Chapter 2, that collisions with the walls then predominate. There was then still some discrepancy between experiment and theory. Twomey and Franklin (1974) carried the process one stage further by allowing for radial variations of steady state and perturbed number density, particle velocity, and electron temperature, with the conclusion that such variations do not greatly modify the dispersion (see Fig. 7.9). The reader is referred to the original papers for the detail of the actual equations and numerical methods of solution, which, while straightforward, would take too much space to reproduce here.

Some discussion is warranted of the appropriate boundary conditions to be applied at the wall and this can be taken in mathematical or in physical terms. Mathematically the form of the equations for the perturbed velocity can be examined; it can be shown generally that if the axial component of perturbed velocity is to remain finite, its radial gradient must be zero, and this in turn requires the radial component of perturbed velocity to be zero. This is most readily seen by considering the electron equation of motion ignoring electron inertia,

$$\nu_e m \mathbf{v}_{e1} = -e\mathbf{E}_1 - \frac{1}{n_0}\nabla(n_1 k_B T_e),$$

which shows that $\mathbf{v}_{e1} = (e/m\nu_e)\nabla\psi_1$, where $\psi_1 = \phi_1 - (n_1 k_B T_e/n_0)$. Ambipolarity gives $\mathbf{v}_{e1} = \mathbf{v}_{i1}$ and thus $\nabla\times\mathbf{v}_{i1} = 0$, which for

$$\mathbf{v}_{i1} = (v_{iz1}\mathbf{k} + v_{ir1}\hat{\mathbf{r}})\exp\{i(-\omega t + kz)\},$$

gives

$$ikv_{ir1} = \frac{dv_{iz1}}{dr}.$$

Physically one can regard the condition as implying that the Bohm condition holds even in the non-steady state, as already remarked in Chapter 5.

7.6. Waves in plasmas with relative drift between ions and electrons

Since in any current-carrying discharge the ions and electrons are in relative motion, real effects arise in experiments which need to be taken into account. The dispersion equations for such a situation are straightforward to write down but are dependent on the frame of reference chosen. The treatment can be in terms of the fluid equations or the kinetic equations and some important differences emerge as will be seen in § 7.8.

If the frame of reference chosen is that in which the ions are at rest and we consider waves with phase velocity high compared to the electron and ion thermal speeds, it is possible to write as an approximation to

$$\varepsilon(\omega, k) = 1 - \frac{\omega_{pe}^2}{k^2 c_e^2} Z'\left(\frac{\omega}{kc_e} - \frac{v_D}{c_e}\right) - \frac{\omega_{pi}^2}{k^2 c_i^2} Z'\left(\frac{\omega}{kc_i}\right),$$

$$\varepsilon \approx 1 - \frac{\omega_{pe}^2}{(\omega - kv_D)^2} - \frac{\omega_{pi}^2}{\omega^2}.$$

Note that the assumptions require $\omega/k > c_e + v_D$ and $\omega/k > c_i$, implying $v_D \gtrsim c_e$.

This dispersion relation can be examined for solutions in complex ω and real k. It is readily seen that for there to be a repeated root corresponding to a branch point, $\partial\varepsilon/\partial\omega = 0$ or $kv_D/\omega = 1 + (M/m)^{\frac{1}{3}}$, and that for any ω therefore there will be a range of k for which there will be instability. The limiting values substituting back in $\varepsilon = 0$ are

$$\omega = \omega_{pi}^{\frac{2}{3}}\omega_{pe}^{\frac{1}{3}}\left\{1 + \left(\frac{m}{M}\right)^{\frac{1}{3}}\right\} \quad \text{and} \quad \frac{kv_D}{\omega_{pe}} = \left\{1 + \left(\frac{m}{M}\right)^{\frac{1}{3}}\right\}^{\frac{2}{3}}. \qquad (7.9)$$

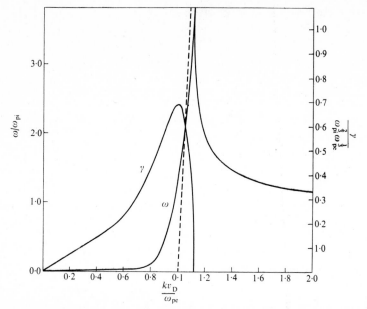

FIG. 7.10. Dispersion and temporal growth-rate curves for the electron–ion two-stream instability in the ion rest frame. Frequency is normalized to the ion plasma frequency ω_{pi} and the growth rate γ to $\omega_{pi}^{\frac{2}{3}}\omega_{pe}^{\frac{1}{3}}$; the wavenumber is normalized by the relative drift v_D and the electron plasma frequency. The electron mode $\omega = kv_D - \omega_{pe}$ is shown dashed.

For k less than this value, ω is imaginary. It is convenient to write the dispersion equation as

$$\frac{\omega - kv_D}{\omega_{pe}} = \frac{\omega}{(\omega^2 - \omega_{pi}^2)^{\frac{1}{2}}}$$

for $\omega > \omega_{pi}$.

The dispersion equation and the growth rate as a function of k are shown in Fig. 7.10. This is a special case of what is usually referred to as the *two-stream* instability. Since the most unstable waves have phase velocity

$$\frac{\omega_r}{k} \approx 0 \cdot 7 v_D \left(\frac{m}{M}\right)^{\frac{1}{3}},$$

we require $v_D > c_e$ for the fluid approximation to the electron and ion susceptibilities to be valid.

To transfer to the electron rest frame of reference, we define $\omega' = \omega - kv_D$, $k' = k$, and then it is seen that the range of frequencies which are

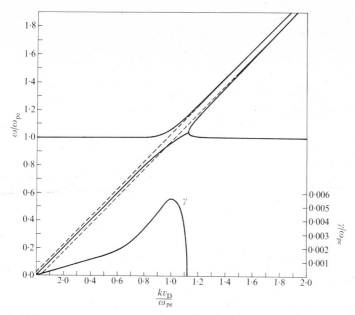

FIG. 7.11. The electron–ion two-stream instability in the electron rest frame with frequency ω and growth rate γ normalized to the electron plasma frequency and the ion beam modes indicated by dashed lines.

unstable is $0 < \omega' < \omega_{pe}\{1 + (m/M)^{\frac{1}{3}}\}$ and the dispersion relation is very closely

$$\frac{\omega'}{\omega_{pe}} = \frac{k v_D}{\omega_{pe}}.$$

The dispersion relation in this frame of reference is given in Fig. 7.11 and is essentially similar to the electron beam/plasma instability. It is also clear that the instability arises from coupling of the slow (negative energy) ion-beam mode with the (positive energy) electron-plasma oscillation (for a discussion of these concepts see Bekefi (1966).

The experimental evidence for the existence of this rapidly growing instability is confined to basic electron-streaming devices (see e.g. Atkinson 1963), although it has been invoked in some gas-discharge situations, particularly where a beam of electrons enters a plasma, e.g. the electrons entering the positive column from the cathode dark space. Measurements in this situation are summarized by Emeleus (1964) (see also Wada, Knechtli, and Heil 1966). In general in gas discharges, however, the drift velocity of electrons is less than the random velocity, and this can be shown theoretically in some specific situations (see e.g. Thonemann 1955; Prewett and Allen (1976), so that it is not possible to attain the drift velocity required and one can legitimately regard the two-stream instability as a collisionless plasma instability. However, plasmas with distributions of particle energies are not stable below the two stream threshold, as will be shown in § 7.8.

7.7. The influence of collisions and finite geometry on the two-stream instability

It is a straightforward matter to include the effect of collisions in the fluid permittivity function we have used to model the two-stream instability in the electron rest frame, to give the relation

$$1 - \frac{\omega_{pe}^2}{\omega(\omega + i\nu_e)} - \frac{\omega_{pe}^2}{(\omega - kv_D)(\omega + i\nu_i - kv_D)} = 0.$$

Thus growth rates are reduced slightly and a threshold introduced, but the modification will not be large unless ν_e, $\nu_i \sim \omega_{pe}^{\frac{1}{3}}\omega_{pi}^{\frac{2}{3}}$.

On the other hand, the influence of finite geometry can be such as to suppress the instability completely. For simplicity, consider the case of a metal cylinder completely filled by a uniform plasma. The dispersion relation in the ion frame is then given by

$$\frac{\omega_{pe}^2}{(\omega - kv_D)^2} + \frac{\omega_{pi}^2}{\omega^2} = 1 + \frac{\beta_{mn}^2}{k^2 a^2} \tag{7.10}$$

in a similar way to the case of ion waves considered earlier in this chapter. For large ω and k we have $\omega - kv_D = \pm\omega_{pe}$ and for large k and small ω, $\omega = \pm\omega_{pi}$ as in the unbounded case but for small ω and k the form is modified so that all four branches may pass through the origin. Approximations

$$\frac{\omega - kv_D}{k} = \pm\frac{\omega_{pe}a}{\beta_{mn}} \quad \text{and} \quad \frac{\omega}{k} = \pm\frac{\omega_{pi}a}{\beta_{mn}}$$

are valid provided that the 'electron' and 'ion' wave dispersion curves do not interact as indicated in Fig. 7.12. The coupling between the waves which gives rise to the instability in a uniform plasma is then absent. The condition for this is readily seen to be

$$v_D - \frac{\omega_{pe}a}{\beta_{mn}} > \frac{\omega_{pi}a}{\beta_{mn}} \quad \text{or} \quad v_D > \frac{c_e}{2^{\frac{1}{2}}\beta_{mn}}\frac{a}{\lambda_D}\left\{1 + \left(\frac{m}{M}\right)^{\frac{1}{2}}\right\}, \tag{7.11}$$

and thus decreases with the order of the mode excited, so that for sufficiently large v_D all modes will be suppressed.

7.8. The ion acoustic instability

At relative drifts less than the electron thermal speed, it is appropriate to use other approximations to the susceptibilities and assuming $\omega/k \ll c_e$,

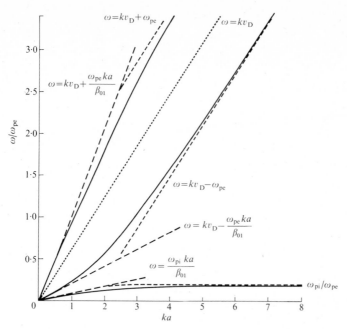

FIG. 7.12. The influence of finite geometry on the two-stream instability indicating that, for sufficiently large v_D, the electron beam modes and the ion plasma mode no longer intersect.

but $\omega/k - v_D \gg c_i$ gives for the real part of ε in the electron rest frame

$$\varepsilon_r = 1 + \frac{2\omega_{pe}^2}{k^2 c_e^2} - \frac{\omega_{pi}^2}{(\omega - kv_D)^2} = 0,$$

or, for small $k\lambda_D$,

$$\frac{\omega - kv_D}{k} = \pm \frac{k_B T_e}{M} \equiv \pm c_s,$$

where c_s is the ion acoustic speed; and for the imaginary part

$$\varepsilon_i = -2\sqrt{\pi}\,\frac{\omega_{pe}^2 \omega}{k^3 c_e^3} - 2\sqrt{\pi}\,\frac{\omega_{pi}^2(\omega - kv_D)}{k^3 c_i^3}\exp\left\{-\left(\frac{\omega - kv_D}{kc_i}\right)^2\right\},$$

and so ω_i/ω_r will change sign as v_D passes through the value v_D^* for which $\varepsilon_i = 0$;

$$v_D^* = c_s\left\{1 + \left(\frac{M}{m}\right)^{\frac{1}{2}}\left(\frac{T_e}{T_i}\right)^{\frac{3}{2}}\exp\left(-\frac{T_e}{2T_i}\right)\right\} \to c_s \quad \text{for} \quad \frac{T_e}{T_i} \gtrsim 25.$$

There will be growth in the appropriate range of frequencies for v_D greater than this value.

General approximate expressions for the dispersion and damping of ion waves with moderate drifts can be shown after some manipulation to be

$$\frac{\omega - k v_D}{k} = \pm c_s (1 + k^2 \lambda_D^2)^{-\frac{1}{2}}$$

and

$$\frac{\gamma}{\omega_{pi}} = \sqrt{\frac{\pi}{8}} \frac{k \lambda_D}{(1 + k^2 \lambda_D^2)^2} \left[\left(\frac{T_e}{T_i}\right)^{\frac{3}{2}} \exp\left\{ -\frac{T_e}{2 T_i (1 + k^2 \lambda_D^2)} \right\} + \right.$$

$$\left. + \left(\frac{m}{M}\right)^{\frac{1}{2}} \left\{ 1 \pm \frac{v_D}{c_s} (1 + k^2 \lambda_D^2)^{\frac{1}{2}} \right\} \right].$$

(7.8a)

Eqns (7.8) are the long-wavelength (small k) limit of these. Thus, for drift velocities significantly above the threshold value for growth, $|\gamma/\omega_{pi}|$ is a function which behaves approximately as

$$\sqrt{\left(\frac{\pi m}{8M}\right)} \frac{v_D}{c_s} \frac{k \lambda_D}{(1 + k^2 \lambda_D^2)^{\frac{3}{2}}}$$

and displays a peak at a wavenumber given approximately by $k \lambda_D \sqrt{2} = 1$, which corresponds in the ion rest frame to a frequency $\omega = \omega_{pi}$. The frequency and growth rate plotted against wavenumber for $T_i/T_e = 0$ and selected values of v_D are given in Figs 7.13 (a) and (b), which show the

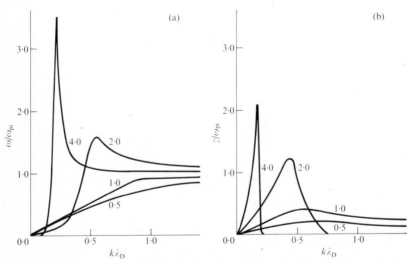

FIG. 7.13. Real and imaginary parts of the frequency for the 'ion acoustic' instability in a plasma with $T_i = 0$, indicating how the dispersion and damping are modified as the relative drift velocity v_D approaches the electron thermal speed c_e, and so the instability becomes essentially two-stream. The parameter is v_D/c_e.

transition from the ion acoustic instability to the two-stream instability with significant increase in growth rates as v_D/c_e increases above unity.

In typical low-pressure discharges, the parameters are such that

$$v_D \equiv \frac{eE}{mv_e} = \mu E = 10^5 \text{ m s}^{-1} \quad \text{for} \quad p = 1 \text{ mTorr,}$$

while the sound speed

$$c_s = \left(\frac{k_B T_e}{M}\right)^{\frac{1}{2}} \approx 10^3 \text{ m s}^{-1}$$

for $T_e \equiv 1 \text{ eV}$ and hydrogen and $c_e = 6 \times 10^5 \text{ m s}^{-1}$, so that the propagation of ion waves is expected to be significantly modified by particle drifts.

It has been suggested that the two-stream instability threshold might set a limiting value on the relative drift and hence on currents which can be drawn through low-pressure discharges, and indeed the ratio of drift velocity to random velocity v_D/c_e does seem experimentally to have a limiting value, but it is typically ~ 0.4 corresponding to the reciprocal of the limit in eqn (4.19) and therefore is not consistent with the value $v_D/c_e \gtrsim 1.34$ required for the onset of the two-stream instability. Such values of axial drift are, however, sufficient to excite the ion acoustic instability, and experimental measurements of the noise emission of positive columns in the frequency range below ω_{pi} have been made by several workers with a view to studying it, e.g. Arunasalam and Brown (1965), Tanaca, Hirose, and Koganei (1967), Fenneman, Raether, and Yamada (1973), and Ilic, Wheeler, Crawford, and Self (1974). The most recent in comparing with theory takes the effect of collisions with the neutral gas into account through a Bhatnagar–Gross–Krook collision term and includes finite ion and electron temperature based on the steady-state theory of the column of Ilic (see p. 86).

Fig. 7.14 gives a comparison of the predicted spatial growth rate for the instability and the observed spectrum of fluctuations in the electron density, where it can be seen the good qualitative agreement is found and indeed the principal features are adequately described.

7.9. Ion waves propagating across the positive column

In general it is possible for ion waves to propagate at any angle to the column axis, and there has been interest in radially propagating ion waves in the same way as with electron waves considered in the previous chapter.

Since the ion acoustic speed c_s is an important parameter in ion wave propagation, and since, as we saw in Chapters 2 and 4, the ion radial velocity passes through the value c_s as the ions pass from plasma to sheath, it is to be expected that ion waves would undergo an important

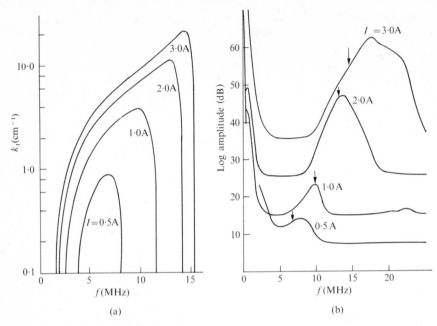

FIG. 7.14. (a) Calculated spatial growth rates as a function of frequency for the ion acoustic instability as a function of discharge currents under conditions appropriate to helium with $pr_w = 2.5$ mm Torr. (b) Stationary spectra observed as fluctuations in the electron density in the appropriate frequency interval. The curves have relatively displaced reference level on the amplitude scale. Arrows mark the maxima of (a).

modification as they propagate into the sheath. This problem and the related one of the appropriate boundary condition or reflection coefficient for ion waves have excited considerable interest, especially since the problem is analogous to fluid wave propagation under trans-sonic conditions. The conclusion of Cavaliere, Engelmann, and Sestero (1970) is that there is no reflection, i.e. there is complete absorption of outward propagating ion waves.

Experiments aimed at testing these conclusions have been performed by Goldan and Leavens (1970) and compared with theoretical values for the phase velocity of ion waves in the sheath. Fig. 7.15 shows this comparison indicating that the phase velocity goes to zero at a distance 10–15 Debye lengths in front of a wall. They also attribute the large voltages needed to launch disturbances from a floating probe to the need for the sheath disturbance to reach beyond this turnover point.

A clear summary of experimental results and exposition of a slab model for radially propagating waves has been given by Weynants, Messaien, and Vandenplas (1973). They specifically include in the steady state and the perturbed equations the effects of inequality of space-charge density

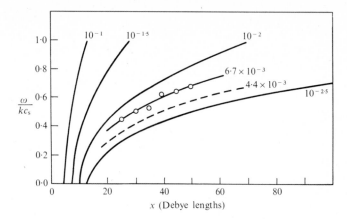

FIG. 7.15. Comparison between theoretical and experimental phase velocity normalized to the ion acoustic speed for radially propagating waves close to the wall of a discharge. The theoretical curves are for different values of the parameter Debye length/discharge radius. The data were measured in argon at 0·33 mTorr. The distance from the wall x is measured in Debye lengths.

and the radial ion velocity. This results in an equation for the potential ϕ which, introducing the normalized ion velocity u, is

$$u^2 \frac{d^4\phi}{dx^4} - 2iu\frac{\omega}{c_s}\frac{d^3\phi}{dx^3} - \frac{\omega^2}{c_s^2}\frac{d^2\phi}{dx^2} +$$

$$+ \frac{1}{\lambda_D^2}\left\{(1-u^2)\frac{d^2\phi}{dx^2} + 2iu\frac{\omega}{c_s}\frac{d\phi}{dx} + \frac{\omega^2}{c_s^2}\phi\right\} = 0. \qquad (7.12)$$

This equation reduces for a uniform long-wavelength situation to two beam modes, $\omega/kc_s = u \pm 1$, but there are additional modes given approximately by

$$k = \frac{\omega}{u} \pm \left(\frac{1-u^2}{u^2}\right)^{\frac{1}{2}}\frac{1}{\lambda_D}$$

and these exist as undamped waves only in the plasma region. Thus even in this model 'information' is not carried through the sheath away from the wall. Eqn (7.12) is, of course, similar to that found for electron waves (6.9) except that the analogous condition to the wave frequency being equal to the local plasma frequency is that the radial ion velocity reaches the sound speed and thus one finds approximately equally spaced resonant frequencies. Solutions to (7.12) have been found for the case ϕ odd and fluctuation current zero at the walls, and the characteristic impedance has been determined as a function of frequency. This leads to some correction to the simple model of the resonant frequencies being given by

$$f = \frac{(2n-1)c_s}{4r_w},$$

corresponding to waves of phase velocity c_s being reflected from the walls of the discharge and symmetric modes can be treated within the same framework. The correction factors are of the order of 1, and interestingly, if we compare with the plasma balance eqn (2.4), allow the ionization rate to be estimated by an alternative method to that described in § 5.12.

8

IONIZATION WAVES

8.1. Non-isothermal waves at low pressures

THE inclusion of variation of electron temperature in the perturbations describing wave propagation requires the electron energy equation to be included in the set of conservation equations. Following eqn (1.15) this can be written

$$\frac{3}{2}\frac{\partial}{\partial t}(nk_B T_e) + \tfrac{5}{2}\nabla \cdot (nk_B T_e v_e) = -en\mathbf{E} \cdot v_e - \kappa\nu_e nk_B T_e - \frac{\partial}{\partial t}\left(\frac{mnv_e^2}{2}\right).$$

The continuity equation must now take account of the fact that T_e is of the form $T_{e0} + T_{e1}\exp\{i(-\omega t + kz)\}$ and similarly for n and v_{ez}, so that the perturbed equations are

$$-i\omega n_1 + ik(n_1 v_{e0} + n_0 v_{e1}) = n_0 Z' T_{e1} + n_1 Z \qquad (8.1)$$

and

$$-i\omega n_1 + ik(n_1 v_{i0} + n_0 v_{i1}) = n_0 Z' T_{e1} + n_1 Z, \qquad (8.2)$$

where $Z' \equiv (dZ/dT_e)_{T_{e0}}$.

The energy equation gives in the steady state $eE_0 v_{e0} = -\kappa\nu_e k_B T_{e0}$, and in the perturbed state

$$-\tfrac{3}{2}i\omega(n_0 k_B T_{e1} + n_1 k_B T_{e0}) + \tfrac{5}{2}ik(n_1 k_B T_{e0} v_{e0} + n_0 k_B T_{e1} v_{e0} + n_0 k_B T_{e0} v_{e1})$$
$$= -e(n_0 v_{e0} E_1 + n_1 v_{e0} E_0 + n_0 v_{e1} E_0) - \kappa\nu_e k_B(n_1 T_{e0} + n_0 T_{e1}) \quad (8.3)$$

where it has been assumed that the electron drift energy is small compared with the thermal energy and the collision frequency is large compared with the ionization rate. In order to understand the types of wave involved, it is necessary to make some simplifying assumptions and for the case where the variables are ordered according to $kv_{e0} \gg \omega$, $Z'T_{e1} \gg Z$, kv_{i0}, the equations reduce to

$$ik(n_1 v_{e0} + n_0 v_{e1}) = 0, \qquad \text{electron continuity;}$$

$$-i\omega n_1 = n_0 Z' T_{e1}, \qquad \text{ion continuity;}$$

$$eE_1 = -ikk_B\left(T_{e1} + T_{e0}\frac{n_1}{n_0}\right), \qquad \text{electron momentum;}$$

$$-\tfrac{5}{2}ik\frac{k_B T_{e1}}{eE_0} = \frac{E_1}{E_0} - \frac{T_{e1}}{T_{e0}} - \frac{n_1}{n_0}, \qquad \text{electron energy;}$$

or

$$\frac{-i\omega(n_1/n_0)}{Z'T_{e0}} = \frac{T_{e1}}{T_{e0}} = -\frac{eE_1}{ikk_B T_{e0}[1-(Z'T_{e0}/i\omega)]},$$

with a dispersion relation given by the determinant.

Introducing the dimensionless variables

$$W = \frac{\omega}{Z}, \qquad H = \frac{T_{e0}Z'}{Z}, \quad \text{and} \quad K = k\frac{k_B T_{e0}}{eE} = kL$$

gives the dispersion relation

$$iW = \tfrac{2}{3}H\left(\frac{1+iK}{\tfrac{2}{3}-iK}\right) \equiv -\tfrac{2}{3}H + \left(\frac{10H/9}{\tfrac{2}{3}-iK}\right), \qquad (8.4)$$

the detailed form of which is dependent on whether ω or k is taken to be complex because the damping is large in some regions. If we write $k = k_r + ik_i$, corresponding to spatial variation, then it is found that

$$k_r L = \frac{10HW}{9W^2 + 4H^2}, \qquad k_i L = \frac{2}{3}\left(\frac{6H^2 - 9W^2}{4H^2 + 9W^2}\right),$$

so that for $W \gg H$, $k_r L \sim 10H/9W$ and $k_i L \sim \tfrac{2}{3}$, implying a backward propagating wave in which the electron thermal fluctuation dominates. Such a wave is clearly physically a new type of wave which we shall identify as an ionization wave. On the other hand, if $W = W_r + iW_i$, corresponding to temporal variation, then the dispersion relation is

$$W_r = \frac{10kLH}{4 + 9k^2 L^2}, \qquad W_i = -H\left(\frac{4 - 6k^2 L^2}{4 + 9k^2 L^2}\right),$$

which for large kL gives

$$W_r \sim \frac{10H}{9kL} \quad \text{and} \quad W_i \sim \tfrac{2}{3}H.$$

Both relations are shown in Fig. 8.1 where the frequency scale is modified by defining $W' = W/H \equiv \omega/T_{e0}Z'$. The complex ω curve however, is, not consistent with the assumptions made in deriving it, and so there is a need for a more general treatment.

Also implicit in the above treatment has been the inclusion of the relaxation time τ_T and length λ_T for electron temperature. From the energy equation these are readily found:

(1) by setting $\mathbf{v}_e = 0$ we obtain $\tau_T = \tfrac{3}{2}\kappa\nu_e$;

(2) by setting $\mathbf{v}_e = 0$ while taking $n\mathbf{v}_e$ constant, corresponding to deceleration of a steady current of electrons while undergoing collisions, we obtain $\lambda_T = \tfrac{3}{2}|v_{e0}|/\kappa\nu_e$ and together with the steady-state energy equation $ev_{e0}E = -\kappa\nu_e k_B T_e$ this gives $\lambda_T = \tfrac{3}{2}(k_B T_e/eE_0)$.

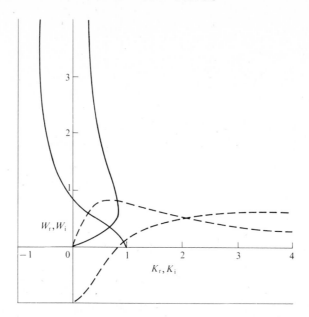

Fɪɢ. 8.1. Dispersion relations for ionization waves on a simple model for real ω and complex k (full curves) and for complex ω and real k (dashed curves). Frequency is normalized to the logarithmic temperature derivative of the ionization rate $T_{e0}\, dZ/dT_{e0}$ and the wavenumber to $k_B T_{e0}/eE_0$, E_0 being the longitudinal field.

More detailed treatments give results which differ only by a numerical factor close to unity (see e.g. Granowski 1955). For comparison note that the steady-state momentum equation is $v_{e0} = -[eE_0/(\nu_e + Z)m]$, so that λ_T can be written $\frac{3}{2}\{[eE_0/m\kappa\nu_e(\nu_e + Z)]\}$ and the equivalent mean free path model would give

$$\frac{3}{4}\left\{\frac{eE_0}{k_B T_e \kappa[1+(Z/\nu_e)]}\right\}.$$

The normalization of wavenumber made above then is effectively to the electron-temperature relaxation length.

8.2. The relation between ion waves and ionization waves

The retention of ion inertia in the equations of the preceding section renders them necessarily more cumbersome, but with some manipulation they can be shown to give a cubic dispersion equation in K and a quadratic in W, with the additional root being identifiable as the ion wave at high frequencies and large wavenumbers and corresponds to $T_{e1} \to 0$. Analyses along these lines have been given by Swain and Brown (1971) and Franklin (1970) retaining ion drift, ion collisions, and thermal relaxation, allowing for the fact that λ_T becomes significantly greater then $k_B T_e / e E_0$ at low pressures.

The dispersion equation is then

$$\left(ik - \frac{1}{\lambda_T}\right)\Omega^2 + i\Omega\left\{(1+\delta_i)\left(ik - \frac{1}{\lambda_T}\right) - \tfrac{2}{3}H\left(\frac{2}{L} + ik\right)\right\} +$$

$$+ \frac{2}{3}\frac{ikL}{D^2}\left\{(1+ikL)\left(\frac{5}{2}ik - \frac{1}{\lambda_T}\right) + ik\right\} +$$

$$+ \frac{2}{3}H(2+ikL)\left\{\frac{(1+\delta_i)}{L} + \frac{ik}{(1+\delta_i)D^2}\right\} = 0,$$

where

$$\Omega = \frac{\omega - kv_{i0}}{Z}, \qquad D^2 = \frac{k_B T_{e0} Z^2 M}{e^2 E^2}$$

and the other symbols are as previously defined.

For λ_T infinite, H zero, one finds a large k large ω solution which is $\omega^2 = k^2 . \tfrac{5}{3}c_s^2$, i.e. adiabatic ion waves, while for small k and large ω a modified form of eqn (8.4) results.

The measure of agreement which can be achieved between theory and experiment at low pressure is shown in Fig. 8.2, where the results of Barrett and Little (1965) for backward waves in mercury are compared with calculations for the appropriate parameters.

The details of the coupling between ion and ionization waves are dependent on the gas parameters and particularly that which we have called H, the variation of ionization rate with temperature, as can be seen in Fig. 8.3. For a linear increase in the ionization cross-section with energy above threshold and a Maxwellian distribution function, i.e. Z given by eqn (2.2), H can be shown to be $\tfrac{3}{2} + (eV_i/k_B T_e)$, where V_i is the ionization potential of the atom or molecule concerned. Of course both the assumptions leading to this value are not universally true and will be discussed later.

8.3. Ionization waves at higher pressures

At higher gas pressures the ion motion is dominated by collisions and the ionization processes in the plasma column are predominantly two-stage. This means that apart from the sort of wave equation given above indicating the existence of backward waves, the balance equations should include the influence of two-stage processes. In the rare gases these will be largely ionization by electron impact with metastable atoms.

Historically this was the regime in which experiments were carried out and the analytical work was much influenced by the physical conditions.

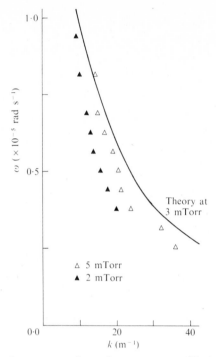

$\omega \ (\times 10^{-5} \ \mathrm{rad \ s^{-1}})$

Theory at
3 mTorr

△ 5 mTorr
▲ 2 mTorr

$k \ (\mathrm{m^{-1}})$

FIG. 8.2. Comparison between experimental measurements of ionization waves in mercury
at low pressure and theory for parameters corresponding to 3 mTorr.

However, the collisional limit can be related to the low-pressure treatment of the first section of this chapter, and indeed, the equation for the real parts of ω and k, namely,

$$\omega k \sim \frac{10}{9} \frac{e E_0 Z'}{k_{\mathrm{B}}},$$

is a simplified form, apart from the numerical factor, of the results of Pekarek and Krejci (1963).

Their treatment was based on a consideration of the response of the plasma to an impulsive disturbance and intended for application at higher pressures where it is more natural to think in terms of diffusion coefficients rather than collision frequencies. This is clearly the more physical approach to the problem, but since the diffusion coefficient and mobility are related by $D_{\mathrm{e}}/\mu_{\mathrm{e}} = k_{\mathrm{B}}T_{\mathrm{e}}/e$ and since $\mu_{\mathrm{e}} = e/m\nu_{\mathrm{e}}$, the equivalence can be shown. They also allowed for the variation of κ with T_{e} in which gives an addition term $\kappa' \nu_{\mathrm{e}} k_{\mathrm{B}} T_{\mathrm{e}1} n_0 T_{\mathrm{e}0}$ in (8.3) and κ', the derivative of κ with respect to T_{e}, may be determined from Fig. 1.5 (p. 7).

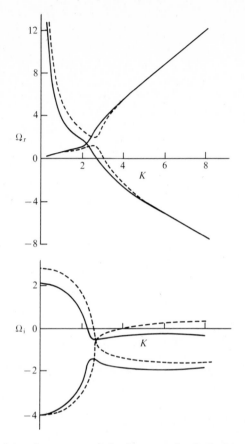

FIG. 8.3. Real and imaginary parts of the frequency for ionization and ion waves for parameters appropriate to argon at a pressure of 10 mTorr showing that coupling between them is a function of the parameter $H \equiv (T_{e0}/Z)(dZ/dT_{e0})$.

They also included the contribution to ionization of two-stage processes. In these circumstances the ionization rate can be written formally as Zn. But it includes terms proportional to the electron density squared at least. If it is written as $Xn + Yn^2$ then the perturbed term becomes $n_0 T_{e1}(X' + Y'n_0) + n_1(X + 2Yn_0)$, and this can be included formally by allowing Z to be a function of n_0 and suitably modifying the definition of H.

With these factors taken into account the relation (8.4) is modified to

$$i\omega = \frac{k_B T_e}{M\nu_i + m\nu_e} k^2 - \tfrac{2}{3}ZH + 10 \frac{ZH}{9L} \bigg/ \left(\frac{1}{\lambda_T} - ik\right); \qquad (8.5)$$

the first term is just $D_a k^2$, where D_a is the ambipolar diffusion coefficient,

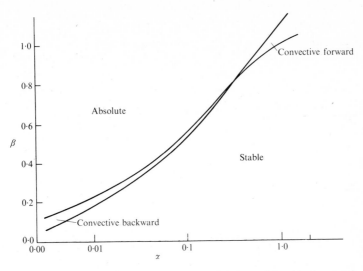

FIG. 8.4. Regions in the space of the parameters α and β determining whether the plasma is stable to ionization waves and the nature of the instability when unstable;

$$\alpha \equiv \frac{2D_a}{3\lambda_T^2 Z'(T_{e0} + E_0\lambda_T)}, \qquad \beta \equiv \left(1 + \frac{E_0\lambda_T}{T_{e0}}\right)^{-1},$$

where λ_T is the electron temperature relaxation length and D_a the ambipolar diffusion coefficient.

and this form of dispersion has been discussed by Garscadden and co-workers (1966, 1969) and related to experimental results. In particular they have given the most comprehensive consideration of the nature of any instability, i.e. whether it is convective or absolute. A convective instability is one in which growth occurs in the wave frame but not in the rest frame, whereas an absolute instability involves growth everywhere. The analysis was carried out for pressures of the order of 1 Torr, and the regions of instability found in terms of parameters α and β defined by

$$\alpha = \frac{D_a}{\frac{3}{2}\lambda_T^2 Z' T_e\{1 + (eE_0\lambda_T/k_B T_e)\}} \quad \text{and} \quad \beta = \frac{1}{1 + (eE_0\lambda_T/k_B T_e)}. \quad (8.6)$$

For the model chosen the second term in the denominator was taken to be $3\pi/16$ making $\beta = 0.63$ and $\alpha = 0.42(D_a Z/\lambda_T^2 H)$. Fig. 8.4 gives the results of computations showing the regions of instability, determined using the colliding-pole and frequency-loop criteria, and indicating that convective instability occurs only for a very limited region of parameter space. However, it has been possible to demonstrate the transition, and Fig. 8.11 (p. 180) shows this for the saturated instability very clearly.

8.4. Metastable-guided ionization waves

The inclusion of two-stage processes in the generation of charged particles, while an important step, gives an incomplete description since the metastable density itself may be modulated by the passage of an ionization wave. One would expect that explicit inclusion of the metastable density in the steady state and perturbed equations would introduce another type of wave.

Pekarek, Masek, and Rohlena (1973) have set up a simplified model which allows the existence of what they have called a metastable-guided wave to be demonstrated.

The model supposes that only one metastable state is significant. The processes included then are:

(1) direct ionization, rate $n_e n_g Z_{gi}$;
(2) excitation to the metastable level, rate $n_e n_g Z_{gm}$;
(3) ionization of metastable atoms, rate $n_e n_m Z_{mi}$;
(4) de-excitation of metastables by collisions of the second kind, rate $n_e n_m Z_{mg}$.

Diffusion is the other loss process and the particle motion is mobility controlled.

The governing equations corresponding to those in § 5.10 would then become as follows.

$$(1) \qquad \frac{\partial n_m}{\partial t} = D_m \nabla^2 n_m + n_e n_g Z_{gm} - n_e n_m (Z_{mi} + Z_{mg})$$

describing metastable generation and loss;

$$(2) \qquad \frac{\partial n_i}{\partial t} + \mathbf{v}_i . \nabla n_i + n_i \nabla . \mathbf{v}_i = D_a \nabla^2 n_i + n_e n_g Z_{gi} + n_e n_m Z_{mi}$$

describing ion generation and loss, where for the steady state the left-hand side is zero and the variation of ambipolar diffusion coefficient D_a with electron mean energy enters through the second and third terms:

$$(3) \qquad M\mathbf{v}_i = \mu_i e \mathbf{E},$$

ion motion mobility determined;

$$(4) \qquad m\mathbf{v}_e = -\mu_e e \mathbf{E} - \frac{1}{n} \nabla(n k_B T_e),$$

electron motion, determined by mobility and electron pressure;

$$(5) \qquad \tfrac{5}{2}\nabla . (n k_B T_e \mathbf{v}_e) = -en\mathbf{E} . \mathbf{v}_e - \kappa \nu_e n k_B T_e$$

gives the electron temperature, variations in time being taken as slow compared to the wave frequency. Introducing the effective transverse discharge dimension Λ we can define $\tau_m = \Lambda^2/D_m$, and $\tau_i = \Lambda^2/D_a$ as lifetimes of the metastables and ions for diffusion. The steady-state equations then become

and
$$n_e n_g Z_{gm} = n_m(\tau_m^{-1} + n_e Z_{mi} + n_e Z_{mg})$$

$$n_i \tau_i^{-1} = n_e(n_g Z_{gi} + n_m Z_{mi})$$

for given Zs, τs, and n_g (or gas pressure), and assuming quasi-neutrality ($n_e = n_i$) the charged-particle and metastable densities can be related.

The equations for the perturbed variables can then be found by linearizing, and a dispersion relation is given by the vanishing of the determinant. In fact Pekarek *et al.* used the mean energy rather than the temperature as a parameter, since under high-pressure conditions the distribution function is more nearly Druyvesteyn than Maxwellian.

The form of the dispersion equation is

$$\omega^2 + i\omega\left(\frac{D_1(k)}{\tau_i} + \frac{A_1(k)}{\tau_m}\right) + \frac{B_1(k)C_1(k) - A_1(k)D_1(k)}{\tau_i \tau_m} = 0,$$

where A_1, B_1, C_1, and D_1 are dependent on the particular gas, with B_1 and C_1 determining the coupling between metastable and ion-density perturbations, while A_1 and D_1 determine the separate characteristics of metastable guided and ion wave modes in the absence of coupling.

Fig. 8.5 gives the calculated dispersion and damping for both waves in

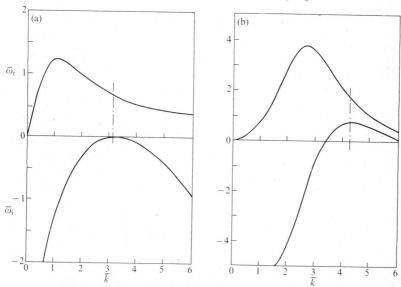

FIG. 8.5. Dispersion and damping curves for ionization waves in a plasma with metastable atoms contributing to the ionization process in neon at pressures of the order of 1 Torr. The curves are normalized to the lifetime of the ions τ_i for the normal ionization wave (a) and to the lifetime of the dominant metastable atom τ_m for metastable waves (b) and the wavenumber by $\bar{\varepsilon}/eE_0$.

TABLE 8.1

Normal-ion-guided wave $E_z = 4.72\text{V cm}^{-1}$, $p = 1.25$ Torr, $I = 8$ mA, $r_w = 8$ mm

	Wavelength λ(cm)	Frequency f(Hz)	Phase velocity v_ϕ(m s^{-1})	group velocity v_g(m s^{-1})
Theory	2·71	1.75×10^4	−504	−1·1
Experiment	2·70	1.59×10^4	−1160	−2·7

Metastable-guided wave $E_z = 4.52\text{V cm}^{-1}$, $p = 1.5$ Torr, $I = 10$ mA, $r_w = 8$ mm

	Wavelength λ(cm)	Frequency f(Hz)	Phase velocity v_ϕ(m s^{-1})	Group velocity v_g(m s^{-1})
Theory	1·98	2.48×10^3	49·6	−143
Experiment	2·08	3.27×10^3	68·0	−425

neon and is in good qualitative agreement with experiment as can be seen from the values given in Table 8.1.

8.5. The influence of electron energy distribution

The details of the inelastic collision processes have a significant effect on the form of the electron energy distribution and in, general, at higher pressures it is not appropriate to use a simple analytical form as we have already seen in relation to the contraction of the positive column.

The problem in seeking to determine the theoretical energy distribution is basically how much information to include in the modified Boltzmann equation and to assess its significance. Most work in this area has had to be numerical.

Swain and Brown (1971), examining the coupling between ion waves and ionization waves, were concerned primarily with modification to the ionization rate and its dependence on mean energy, and they carried out computations for argon at low pressure. A comparison of such an analysis with experiment is shown in Fig. 8.6, in which the agreement with experiment is closer than that obtained with an assumed Maxwellian distribution.

Another example of a treatment which considers the details of the electron distribution function is that of Wojaczek (1971) for argon at high currents and pressures when the body of the distribution can be expected, according to eqn (1.12), to be Maxwellian. Inclusion of the effects of heat conduction and thermal diffusion together with the major inelastic collision processes lead to the agreement between theory and experiment shown in Fig. 8.7.

The most comprehensive treatment of the effects of inelastic collisions on the distribution function is that of Gentle (1966) for argon at medium

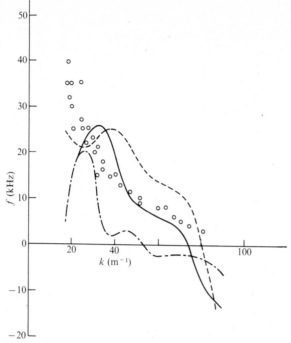

FIG. 8.6. Measured and computed dispersion of ionization waves at low pressure (5 mTorr) in argon. In the computations the parameters have been chosen for the solid line to give a good fit with experiment and the dashed line represents a change in the value of the excitation cross-section used in computing the distribution function; the chain line represents the effect of an increase in pressure to 10 mTorr other parameters held constant.

pressures and currents, and the agreement between theory and his experimental measurements made on standing waves in a flowing discharge, as seen in Fig. 8.8, is remarkable. Unfortunately it is not possible without varying the parameters in purely numerical work to determine what processes are physically important.

8.6. The effect of finite column radius on ionization waves

We have seen that ionization waves arise physically from variations in the ionization rate through the electron temperature. In the steady state the electron temperature is determined by the plasma balance eqn (2.4), which relates ionization and loss rates. These latter depend on radius, and for this reason one might expect the discharge radius to be an important parameter in determining the characteristics of such waves.

It is possible to set up equations in which the radial variation of steady state and perturbed quantities is explicit and to solve the resulting

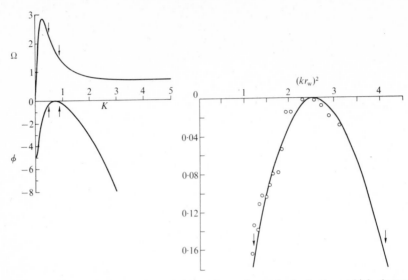

FIG. 8.7. (a) Calculated dispersion relation and growth rate for ionization at high electron densities in argon for a value of $pr_w = 14$ mm Torr. The frequency is normalized to the ion lifetime τ_i and the wavenumber effectively to the ambipolar diffusion coefficient and the ion lifetime, i.e. $K^2 = k^2 D_a \tau_i$. (b) Measured spatial damping rates compared with theory over a range of values near that corresponding to the maximum growth rate γ_0 under the same conditions. The arrows indicate corresponding intervals in the two diagrams.

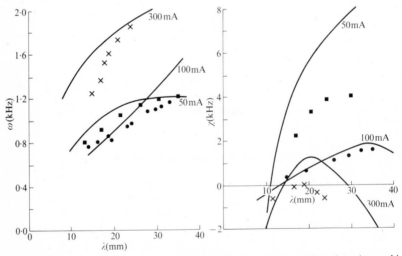

FIG. 8.8. Real and imaginary parts (ω, γ) of the frequency versus wavelength λ observed in argon at 10 Torr, $r_w = 15$ mm, for three different values of mean electron density, i.e. discharge current compared with theoretical computations involving an evaluation of the steady state and oscillating electron energy distribution function including inelastic processes.

equations to determine the dispersion characteristics. Such a treatment in the quasi-neutral cold-ion $(T_i = 0)$ approximation has been given by Twomey and Franklin (1974). The solution is necessarily numerical since one has a set of eight simultaneous non-linear first-order differential equations. Fig. 8.9 gives the results of computations for the ionization wave branch with parameters appropriate to argon at low and medium pressures for the cases in which

(1) radial variations are explicit;
(2) the quantities are assumed radially uniform but with appropriate allowance for particle generation and loss.

It can be seen that the curves are qualitatively very similar but there are some quantitative differences. We conclude that radial variations of the perturbed variables are not important in determining the dispersion.

8.7. Ionization waves in a magnetic field

In a number of experiments in which backward waves have been found and studied, and axial magnetic field was present; often because of the experimental advantage of helping to stabilize the discharge. Since the two-fluid model of the steady state can readily be modified to take account of such a field, it is a natural extension of the theoretical work described earlier in this chapter to include the effect of a magnetic field. This has been done by Duncan and Forrest (1971), who used the device of specifying a wavenumber $k_\perp$ perpendicular to the column axis and determined by the radial boundary conditions to describe the perturbed state. This allows the dispersion equation to be obtained algebraically, albeit with considerable manipulation and has the advantage that the physics of the processes involved can be seen. The resulting equation expressed in the variables used here was

$$K^2[\omega_1\{\omega_1 + i(\nu_i + Z)\} - k^2 c_s^2](\tfrac{3}{2}\omega_2 + i\nu_e + iD_e K^2 - iZg)$$

$$= -i(\nu_i + Z)\frac{\mu_i}{\mu_e}\left(\omega_2 k^2\left\{\tfrac{3}{2}\omega - k_z v_{e0}\left(\frac{7}{2} - \frac{Zg}{\nu_e}\right) + i\nu_e + 2iD_e K^2 + i\frac{Zg}{\kappa}\right\} + $$

$$+ \frac{gZ}{c_s^2}(K^2 D_e + 2ik_z v_{e0})[k_z v_{e0}\{\omega_1 - k_z v_{i0} + i(\nu_i + Z)\} - ik_z v_{i0}(\nu_i + Z) + $$

$$+ k_z^2 v_{i0}^2 - k_z^2 c_s^2]) + \frac{2Zg}{\kappa}\omega_1(K^2 - k_z^2)\{\omega_1 + i(\nu_i + Z)\}, \qquad (8.7)$$

with

$$\omega_1 = \omega - k_z v_{i0}, \qquad \omega_2 = \omega - k_z v_{e0}, \qquad K^2 = k_z^2 + \frac{k_\perp^2 \nu_e^2}{\nu_e^2 + \omega_{ce}^2},$$

$$g = \frac{eV_i}{k_B T_{e0}} + \frac{1}{2} + \left(1 + \frac{eV_i}{2k_B T_{e0}}\right)^{-1}, \qquad \nu_c = \tfrac{3}{2}(Z + \kappa\nu_e) + Zg\frac{eV_i}{k_B T_{e0}} + \tfrac{3}{2}Zg.$$

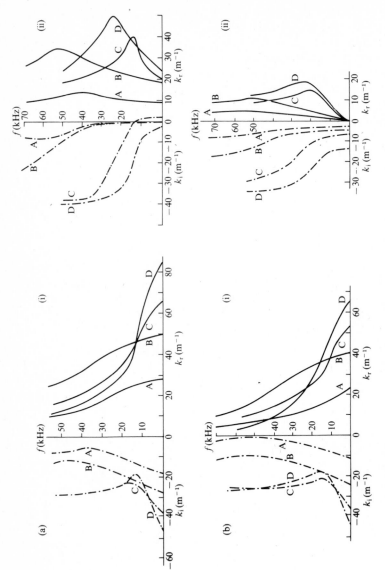

FIG. 8.9. (a) The theoretical influence of radial variation of discharge parameters on the propagation of the fundamental mode of ionization waves: (i) anode-directed group velocity waves, (ii) variably directed wave. (b) Corresponding results for a finite uniform plasma.

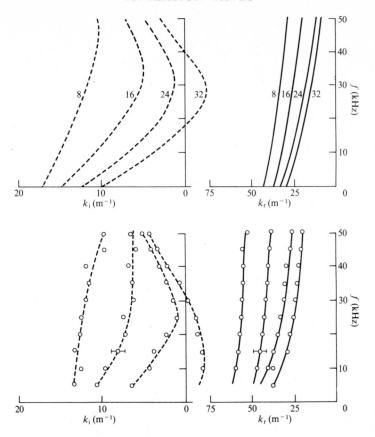

FIG. 8.10. Comparison between experimental results and theory for anode-propagating plane ionization waves in an axial magnetic field. The discharge parameters were: gas—helium; pressure—38 mTorr; radius—12·5 mm; current ≤ 300 mA.

The equation is third-order in ω and sixth-order in k_z, and the second and third factors on the left-hand side correspond to ion waves and ionization waves respectively. The right-hand side represents the coupling between the three types of waves. Fig. 8.10 shows experimental and theoretical dispersion and damping curves compared for a range of magnetic fields in helium at 38 mTorr; qualitatively the principal features are well explained by taking $k \perp = 0$ i.e. plane waves.

8.8. Waves at low pressure with more than one species of ion

If one makes the simple modification of adding a second species of ion to a plasma and calculates the relative permittivity according to kinetic theory it is readily found to be

$$\varepsilon(\omega, k) = 1 - \frac{\omega_{pe}^2}{k^2 c_e^2} Z'\left(\frac{\omega}{kc_e}\right) - \frac{\omega_{pi1}^2}{k^2 c_{i1}^2} Z'\left(\frac{\omega}{kc_{i1}}\right) - \frac{\omega_{pi2}^2}{k^2 c_{i2}^2} Z'\left(\frac{\omega}{kc_{i2}}\right). \quad (8.8)$$

For $\omega/k \ll c_e$ but $\gg c_{i1}, c_{i2}$ this yields the approximate relation

$$1 + \frac{2\omega_{pe}^2}{k^2 c_e^2} = \frac{\omega_{pi1}^2 + \omega_{pi2}^2}{\omega^2}$$

and the ion-wave mode is modified by there being a hybrid ion frequency $\omega_{pi}^{2*} = \omega_{pi1}^2 + \omega_{pi2}^2$. For charge neutrality with both ions positive, corresponding to a mixture of atomic and molecular ions, say, then $n_e = n_{i1} + n_{i2}$ and for $M_{i1} < M_{i2}$

$$\omega_{pe}^2 \frac{m}{M_{i2}} < \omega_{pi}^{*2} < \omega_{pe}^2 \frac{m}{M_{i1}}$$

and the waves have a phase velocity

$$\frac{\omega^2}{k^2} = c_s^{2*} = \frac{\omega_{pi}^{*2} c_s^2}{\omega_{pi}^2},$$

where $c_{s1}^2 > c_s^{2*} > c_{s2}^2$. Alternatively one of the ions may be negative, corresponding to the situation in oxygen described in § 5.3. Then $n_e + n_{i1} = n_{i2}$, and writing

$$\frac{n_{i1}}{n_{e0}} = \frac{A_0}{D_0} = \alpha_n, \qquad \frac{n_{i2}}{n_{e0}} = 1 + \alpha_n$$

the characteristic frequency ω_{pi}^* is given by

$$\omega_{pi}^{2*} = \omega_{pe}^2 \left(\frac{\alpha_n}{M_{i1}} + \frac{1 + \alpha_n}{M_{i2}} \right) m > \omega_{pi2}^2$$

and the phase velocity by

$$\frac{\omega^2}{k^2} = c_s^{2*}, \quad (8.9)$$

where

$$c_s^{2*} > c_{s2}^2 \quad \text{and} \quad \alpha_n(c_{s1}^2 + c_{s2}^2).$$

Thus for simple electrostatic modes under conditions where the fluid approximation is appropriate, the two species do not make their separate identities felt; however, with cyclotron modes, there is a separate cyclotron resonance for each species.

Further, in the case of a *collisionless* plasma with two ion species where one ion is very much lighter than the other, there is the possibility with a small fraction f of light ions for the combined ion acoustic velocity c_s^* to be comparable to the light ion thermal speed c_{i2}, in which case one would expect significant Landau damping on the light ions. The approximations to the plasma dispersion function allow the damping rate to be written for the case $T_e/T_i \gg 1$, $m_{i1}/m_{i2} \gg 1$ when

$$\frac{\omega^2}{k^2 c_{i1}^2} \approx \frac{m_{i2}(1-f)+m_{i1}f}{m_{i2}} \frac{T_e}{2T_{i1}},$$

as

$$\frac{\gamma}{\omega} = \frac{m_{i2}}{m_{i2}(1-f)+m_{i1}f} \left\{ \left(\frac{T_{i1}}{T_e}\right)^{\frac{3}{2}} \left(\frac{m_e}{m_{i1}}\right)^{\frac{1}{2}} + 2f\left(\frac{m_{i2}}{m_{i1}}\right)^{\frac{1}{2}} \exp\left(-\frac{m_{i2}}{m_{i1}}\frac{\omega^2}{k^2 c_{i1}^2}\right) + \right.$$
$$\left. + 2(1-f)\exp\left(-\frac{\omega^2}{k^2 c_{i1}^2}\right) \right\}.$$

This latter can be shown to have a maximum value given approximately by

$$f^* = \left\{ \left(1 + 8\frac{m_{i1}}{m_{i2}}\frac{T_{i1}}{T_e}\right)^{\frac{1}{2}} + 1 \right\} \Big/ 2\frac{m_{i1}}{m_{i2}},$$

which may be a significant fraction of unity. Under such circumstances the ion modes which can exist are not simply related to those in plasmas containing each ion alone, and there can be two modes with comparable damping and differing phase velocities representing modified light and heavy ion waves, the latter not being the mode corresponding to the least damped root for $f = 0$. This situation has been treated theoretically in detail by Fried, White, and Samec (1971) and experiments performed in helium–argon mixtures by Nakamura, Ito, Nakamura, and Itoh (1975).

The situation described by eqn (8.9) can be generalized in the spirit of § 8.1 to include electron temperature variations by perturbing eqns (5.3)–(5.5) and adding the electron energy equation (8.3). Making the assumption that the wave phase velocity is small compared with the electron drift velocity and modelling the wall loss by a volume loss term as was done in § 8.4, with a rate r_i for positive ions and r_e for electrons, and taking it to be zero for negative ions, allows the following dispersion equation to be derived,

$$\omega_1\omega_2\omega_3\omega_4 - \omega_1\omega_2\omega_4(\beta'' + \tfrac{2}{3}\varepsilon'') + \omega_2\omega_3\omega_4(\delta'' + \tfrac{2}{3}\kappa'') -$$
$$- \tfrac{5}{3}\alpha''\omega_1\omega_4 - \tfrac{5}{3}\gamma''\omega_2\omega_3 = 0, \tag{8.10}$$

where

$$\omega_1 = \omega - \mathbf{k}.\mathbf{v}_{n0} + iD_0, \qquad\qquad \omega_2 = \omega - \mathbf{k}.\mathbf{v}_{i0} + i\nu_i,$$

$$\omega_3 = \omega - \mathbf{k}.\mathbf{v}_{i0} + i\frac{Z_0}{1+(A_0/D_0)}, \qquad \omega_4 = \omega - \mathbf{k}.\mathbf{v}_{n0} + i\nu_n,$$

$$\alpha'' = (1+\alpha_n)k^2 c_i^2, \qquad \beta'' = Z_0, \qquad \gamma'' = \alpha_n k^2 c_n^2, \qquad \delta'' = A_0,$$

$$\varepsilon'' = \frac{Z_0 T_e}{1+(A_0/D_0)}\left(\frac{A_0}{D_0}\right)' \equiv Z_0' - \frac{r_i'}{r_i}Z_0, \qquad \kappa'' = A_0 T_e\left(\frac{A_0}{D_0}\right)',$$

the single prime denoting differentiation with respect to electron temperature. This form reduces to a simplified version of that derived in § 8.2, in the case where $v_{n0} = v_{i0} = 0$ and $A = D = 0$, otherwise there is coupling between the wave modes which are essentially ion waves Doppler shifted by the opposing positive and negative ion motion, together with an ionization wave mode and an attachment–detachment mode.

8.9. Waves in recombination–dominated plasmas

At higher pressures recombination in the volume becomes an important loss process. This is due to the fact that the probability of electron–ion recombination is considerably increased in the presence of a third body (neutral gas atom or molecule), since it is then easier to satisfy conservation of energy and momentum.

The charged-particle balance is then given by

$$\partial n_e / \partial t = Z n_e - \rho_i n_e n_i = 0 \quad \text{with} \quad n_e = n_i,$$

and since ρ_i is relatively independent of electron temperature, the temperature becomes dependent on charged particle density

$$Z(T_e) = \rho_i n_e.$$

The relation between longitudinal field and electron temperature is

$$I_z \cdot E_z = n_e \kappa \nu_e k_B T_e \equiv n_e \nu_u k_B T_e.$$

Now we consider waves with all quantities q varying as $q_0 + q_1 \exp\{i(kz - \omega t)\}$ and make a number of approximations which are valid at higher pressures when recombination is important. First we assume equality of perturbed particle densities, which, as we have seen earlier, implies wavelengths that are long compared with the Debye length. Secondly, electron momentum-transfer collisions are so frequent and perturbations are sufficiently small so that

$$\frac{E_{z1}}{E_{z0}} = \frac{v_{e1}}{v_{e0}}.$$

Thirdly, consistent with the first approximation, $\nabla \cdot \mathbf{I}_{z1} = 0$, thus

$$\frac{n_{e1}}{n_{e0}} + \frac{v_{e1}}{v_{e0}} = 0.$$

With these assumptions the perturbed energy equation yields

$$2 \frac{n_{e1}}{n_{e0}} + \frac{T_{e1}}{T_{e0}} \left(1 + \frac{\nu'_{u0}}{\nu_{u0}} T_{e0} \right) = 0. \tag{8.11}$$

The perturbed equation for electron generation and loss gives

$$-i\omega n_{e1} = -Z n_{e1} + Z' T_{e0} n_{e0} \frac{T_{e1}}{T_{e0}} - \rho_i' T_{e0} n_{e0}^2 \frac{T_{e1}}{T_{e0}}$$

or, rearranging,

$$\left(-\frac{i\omega + Z}{Z}\right) \frac{n_{e1}}{n_{e0}} = \left(\frac{Z' T_{e0}}{Z} - \frac{\rho_i' T_{e0}}{\rho_i}\right) \frac{T_{e1}}{T_{e0}}.$$

Combining with eqn (8.11) gives

$$\omega_r = 0, \qquad \frac{\omega_i}{Z} = -1 - 2\left(\frac{Z' T_{e0}}{Z} - \frac{\rho_i' T_{e0}}{\rho_i}\right) \Big/ \left(1 + \frac{\nu_u' T_{e0}}{\nu_u}\right),$$

thus there is damping $\omega_i < 0$ given that $Z'/Z > \rho_i'/\rho_i$, i.e. Z is a strongly increasing function of T_e and ρ_i is almost independent of it. This is in contrast to the situation for a diffusion-dominated plasma (§ 8.2), and it is interesting to compare the two cases.

Writing the electron generation and loss equation

$$\frac{\partial n_e}{\partial t} = Z n_e + D_a \nabla^2 n_e$$

in the steady state $Z = D_a/\Lambda^2$, where Λ is the characteristic transverse diffusion length. The perturbed equation is

$$-i\omega n_{e1} = -Z n_{e1} + T_{e0} Z' n_{e0} \frac{T_{e0}}{T_{e1}} - D_a k^2 n_{e1},$$

giving

$$\omega_i = Z - \left(2Z' T_{e0}/1 + \frac{\nu_u}{\nu_u} T_{e0}\right) - k^2 D_a$$

which for sufficiently small k and for

$$\frac{1}{T_{e0}} + \frac{\nu_u'}{\nu_u} > 2\frac{Z'}{Z},$$

$\omega_i > 0$. The difference between the two cases arises from the fact that the stabilizing change of recombination rate with number density is twice the destabilizing change for generation by electron impact.

8.10. Waves at high pressures in electron-attaching gases

The development of high-pressure gas lasers in mixtures of gases which give rise to electron-attaching atoms or radicals as a result of dissociation and chemical reactions has led to a need to understand the conditions for stability of such discharges. A comprehensive treatment of the general problem has been given by Haas (1973) and more specifically by Nighan and Wiegand (1974). By combining the results of earlier sections of this chapter, it is possible to recover their principal results. For simplicity only one type of negative ion density n_n and one type of positive ion density n_i is included in the analysis. The electron generation is assumed to occur by ionization (rate, Zn_e) and detachment (rate, Dn_n) and loss by recombination (rate, $\rho_i n_e n_i$) and attachment (rate, An_e). Thus

$$\frac{\partial n_e}{\partial t} = Zn_e + Dn_n - \rho_i n_e n_i - An_e. \qquad (8.12)$$

For negative ions the negative and positive ion recombination rate is $\rho_n n_n n_i$ to give a balance equation

$$\frac{\partial n_n}{\partial t} = An_e - Dn_n - \rho_n n_n n_i. \qquad (8.13)$$

The rates of processes in general will be a summation over more than one specific reaction. If charge neutrality is assumed throughout, then

$$n_i = n_e + n_n \qquad (8.14)$$

and the balance equation is for positive ions, the sum of (8.12) and (8.13). The electron energy equation and its perturbed form deduced in § 8.9 still apply.

The steady-state equations are non-linear in the densities. Given the electron temperature to set coefficients Z, A, D, ρ_i, ρ_n eqns (8.12), (8.13), and (8.14) can be solved simultaneously for n_e, n_n, and n_i.

The perturbed equations are

$$-i\omega n_{e1} = Zn_{e1} + Dn_{n1} - \rho_i n_{e1} n_{i0} - \rho_i n_{e0} n_{i1} - An_{e1} +$$
$$+ Z' n_{e0} T_{e1} + D' n_{n0} T_{e1} - \rho_i' T_{e1} n_{e0} n_{i0} - A' n_{e0} \qquad (8.15)$$

$$-i\omega n_{n1} = An_{e1} - Dn_{n1} - \rho_n n_{n1} n_{i0} - \rho_n n_{n0} n_{i1} +$$
$$+ A' n_{e0} T_{e1} - D' n_{n0} T_{e1} - \rho_n' n_{n0} n_{i0} T_{e1}. \qquad (8.16)$$

The behaviour of the solution is given by the determinant of eqns (8.11), (8.15), and (8.16), which is quadratic in ω corresponding to an ionization wave, and another which can be conveniently called a negative ion wave. Writing the equations formally

$$(-i\omega + \Omega_e)\tilde{n}_{e1} + \alpha\tilde{n}_{n1} + \beta\tilde{T}_{e1} = 0,$$
$$\gamma\tilde{n}_{e1} + (-i\omega + \Omega_n)\tilde{n}_{n1} + \delta\tilde{T}_{e1} = 0,$$
$$\tilde{n}_{e1} + \tilde{n}_{n1} + \varepsilon\tilde{T}_{e1} = 0,$$

we find

$$\varepsilon\{(-i\omega + \Omega_e)(-i\omega + \Omega_n) - \alpha\gamma\} + \alpha\delta - \beta(-i\omega + \Omega_n) = 0,$$

which can be written even more formally

$$\omega^2 - ip\omega + q = 0$$

and has the solution $\omega_r = 0$,

$$\omega_i = \frac{p \pm \sqrt{(p^2 + 4q)}}{2}.$$

Both modes are unstable if $p > 0$ and $-p^2/4 < q < 0$, while one is unstable when $q > 0$.

Now let us write out p and q in terms of physical variables,

$$p = \varepsilon T_{e0} Z'\left(1 - \frac{A'}{Z'}\right) + \left(n_{e0}\rho_i + \frac{n_{n0}}{n_{e0}} D + n_{n0}\rho_n + \frac{n_{e0}}{n_{n0}} A\right)$$

$$q = \varepsilon T_{e0} Z'\left\{\frac{A'}{Z'}(n_{e0}\rho_i + n_{i0}\rho_n + n_{n0}\rho_n) - n_{n0}\rho_n - \frac{n_{e0}}{n_{n0}} A\right\} -$$

$$- \left(\frac{n_{e0}}{n_{n0}} A\rho_i n_{i0} + \frac{n_{n0}}{n_{e0}} \rho_n D n_{i0}\right).$$

Under conditions of interest in self-sustained discharges $q < 0$ since $A' < Z'$ and $p > 0$ so that both modes are unstable, and detailed examination shows that the mode which reduces to the ionization wave in the absence of negative ions is the faster growing, so that it is expected to dominate.

In laser discharges there is sometimes an external source of ionization usually in the form of an electron beam, and a discussion of how this factor can suppress the instabilities described above has been given by Haas (1973).

8.11. Unstable ionization waves and striations

Several of the dispersion relations given earlier in this chapter have shown a transition from damping to growth of ionization waves as the frequency or wavenumber is varied. This implies instability for some values of ω and k and under appropriate conditions, one would expect there to be self-sustained oscillations. It would depend on the magnitude of the variations of the variables in such oscillations whether one would expect a close relation to the linear theories described. In particular the processes limiting growth can exercise an important influence on the spectrum observed.

This said, the temptation is great to use the results of linear theories to explain the readily visible moving and standing striations which occur in most gas discharges and have been catalogued and studied for many years now. In fact the use of linear results has been surprisingly successful, and this has been helped by the development of the oscillographic technique, analogous to that used in television, of modulating the brightness detected by a photomultiplier whose position determines the Y-deflection and whose time-base is in synchronism with the waves.

This allows the propagation of striations to be displayed and recorded. Fig. 8.11 gives an example in a mercury–argon mixture and shows the transition from convective to absolute instability when the discharge current is reduced from 120 mA (high) to 30 mA (low).

A general map of the state of positive columns for the cases of neon, helium, and argon is reproduced from Pekarek (1971) as Fig. 8.12.

A discussion of the evolution of ionization waves using a general model for the amplitude A which contains terms which mathematically ensure

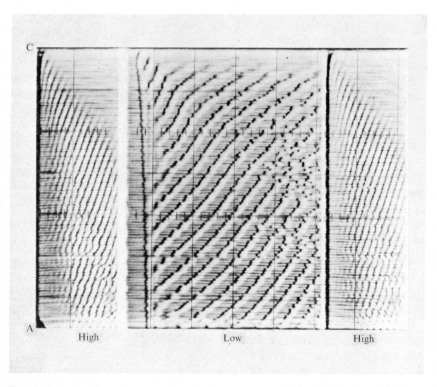

FIG. 8.11. Space–time display of ionization waves in a mercury–argon discharge showing at high currents (120 mA) a convective instability propagating from cathode to anode and at low currents (30 mA) an absolute instability throughout the discharge.

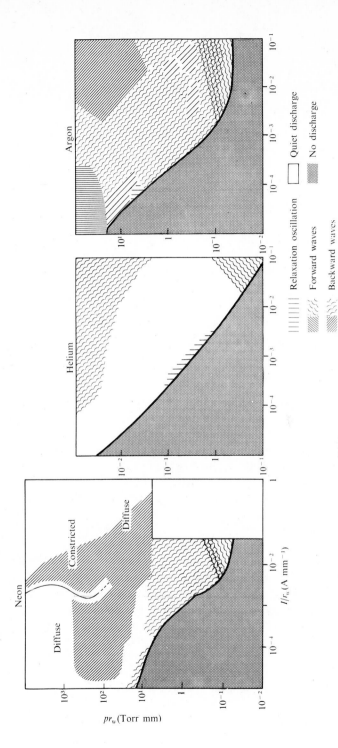

FIG. 8.12. Maps of the behaviour of positive columns in rare gases indicating the nature of spontaneous oscillations observed in space defined by the similarity parameters pr_w and I/r_w.

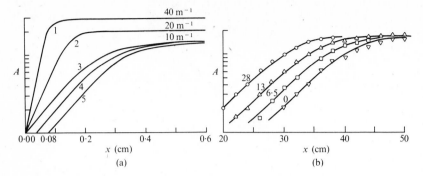

F IG. 8.13. Saturation of an instability (a) modelled by eqn (8.17) with different values of γ (1–3) and different values of x_1 (3–5); (b) experimental measurements with different driving voltages exciting unstable ionization waves at 785 Hz in xenon $pr_w = 9.5$ mm Torr, $I = 0.2975$ A. Compared with theory the parameters A_1 and γ being determined from experiments.

saturation has been given by Sato (1973), who supposed that the evolution was spatial and governed by

$$\mathrm{d}A/\mathrm{d}x = 2\gamma A^2 - \alpha' A^4 - \beta' A^6, \tag{8.17}$$

where γ was the linear growth rate. The problem is to find a physical origin for α' and β'. With such a model the saturation amplitude of A for $\beta'\gamma < \alpha^2/8$ is $\sim(2\gamma/\alpha')^{\frac{1}{2}}$, and with $A(x)$ of the form $A_1 \exp\{\gamma(x - x_1)\}$ for small A gives curves of the form of Fig. 8.13(a) for different values of γ (curves 1, 2, 3) and different values of x_1 (curves 3, 4, 5). Experimental curves of the spatial development with an external exciting voltage applied to a grid in the column at the frequency of the self-excited waves is shown in Fig. 8.13(b). Clearly the proposed equation gives a good description of the behaviour, but the relation of α' and β' to physical quantities has yet to be achieved. γ was found to be a function of the discharge current, and it was shown that γ decreased to zero at a reduced current I/r_w, which coincides with the Pupp (1935) limit above which spontaneous ionization waves do not occur.

   Another interesting departure in the study of the fully developed striations has been given by Grabec and Mikac (1974) who has sought to model the striation as an entity and to examine the nature of the interaction between striations somewhat along the lines of the behaviour of solitons (which have been observed as large amplitude solitary waves in the case of both electron plasma waves and ion waves). Further, under conditions where striations interact, the development of turbulence can be modelled.

9

LOW FREQUENCY WAVES IN MAGNETIZED PLASMA

WITH each of the types of wave considered so far we have dealt with propagation in a uniform plasma first, and then subsequently, treated the effects of radial inhomogeneity and axial magnetic field. However, this sequence was not carried through for ion waves. This is because of the existence of a new wave type in inhomogeneous plasmas at low frequencies, typically of the order of ion plasma and ion cyclotron frequencies. We shall deal with these drift waves in Chapter 10, but before we do so and in order to understand them fully, it is necessary to consider what transverse waves can propagate in a magnetized plasma at low frequencies.

9.1. Transverse wave propagation along a steady magnetic field

We consider a wave with its electric field

$$\mathbf{E} = (E_x \mathbf{i} + E_y \mathbf{j}) \exp\{i(-\omega t + k_z z)\}$$

perpendicular to the magnetic field $\mathbf{B} = B_0 \mathbf{k}$.

The electron motion in the absence of collisions is given by

$$m \frac{d\mathbf{v}_e}{dt} = -e(\mathbf{E} + \mathbf{v}_e \times \mathbf{B}_0)$$

and the ion equation

$$M \frac{d\mathbf{v}_i}{dt} = e(\mathbf{E} + \mathbf{v}_i \times \mathbf{B}_0).$$

Solving for $\mathbf{v}_e$ and $\mathbf{v}_i$ allows one to find $\mathbf{J} \equiv ne(\mathbf{v}_i - \mathbf{v}_e)$ and then the waves are given as solution of Maxwell's equations

$$\nabla \times \mathbf{H} = \varepsilon_0 \frac{\partial \mathbf{E}}{\partial t} + \mathbf{J}, \qquad \nabla \times \mathbf{E} = -\mu_0 \frac{\partial \mathbf{H}}{\partial t}.$$

As is shown in Appendix II (p. 224), this is most conveniently carried out in terms of circularly polarized fields $E_x + iE_y$ and $E_x - iE_y$ to give the dispersion relation

$$0 = -\frac{c^2 k_z^2}{\omega^2} + 1 - \frac{\omega_{pi}^2}{\omega(\omega + \omega_{ci})} - \frac{\omega_{pe}^2}{\omega(\omega - \omega_{ce})} \equiv \varepsilon_R \qquad (9.1)$$

for rotation with the electrons, while the other sense of rotation of the electric field vector gives

$$0 = -\frac{c^2 k_z^2}{\omega^2} + 1 - \frac{\omega_{pi}^2}{\omega(\omega - \omega_{ci})} - \frac{\omega_{pe}^2}{\omega(\omega + \omega_{ce})} \equiv \varepsilon_L. \qquad (9.2)$$

These solutions correspond to the $E_z = H_z = 0$ case in Appendix III (p. 229), where (AIII.17) and (AIII.18) are seen to give $\varepsilon_l = k^2$ and $\varepsilon_r = k^2$, identical with (9.1) and (9.2).

Let us consider limiting cases. For the upper equation, i.e. rotation in the same sense as electrons, there is a resonance ($k_z \to \infty$) and a cut-off ($k_z \to 0$) near $\omega(\omega - \omega_{ce}) = \omega_{pe}^2$, i.e. $\omega = \omega_{ce} + \omega_{pe}^2/\omega_{ce}$ for $\omega_{ce} \gg \omega_{pe}$ or $\omega = \omega_{pe} + \frac{1}{2}\omega_{ce}$ for $\omega_{pe} \gg \omega_{ce}$, and the phase velocity at frequencies $\omega \ll \omega_{ci} \ll \omega_{ce}$ is given by

$$\frac{c^2 k_z^2}{\omega^2} = 1 - \frac{\omega_{pi}^2(\omega - \omega_{ce}) + \omega_{pe}^2(\omega + \omega_{ci})}{\omega(\omega - \omega_{ce})(\omega + \omega_{ci})}$$

$$= 1 + \frac{\omega_{pe}^2 + \omega_{pi}^2}{(\omega_{ce} - \omega)(\omega_{ci} + \omega)}$$

$$\simeq 1 + \frac{\omega_{pe}^2 + \omega_{pi}^2}{\omega_{ce}\omega_{ci}} = 1 + \frac{n}{\varepsilon_0}\frac{(m + M)}{B^2}, \qquad (9.3)$$

which may be written

$$v_\phi^2 = \frac{\omega^2}{k_z^2} = \frac{c^2}{1 + (\rho/\varepsilon_0 B^2)} \sim \frac{B^2}{\mu_0 \rho},$$

where ρ is the total particle mass density. For the other sense of rotation one finds resonance at $\omega = \omega_{ci}$ and a cutoff at $\omega \simeq \omega_{pe}^2/\omega_{ce}$ or $\omega_{pe} - \frac{1}{2}\omega_{ce}$ as $\omega_{ce} \gtrless \omega_{pe}$; the phase velocity at low frequencies tends to the same value.

These limiting cases allow one to draw the dispersion diagram shown in Fig. 9.1, in which logarithmic scales are used because of the fact that several orders of magnitude are involved. The limiting speed of the low-frequency branches is usually referred to as the Alfvén speed $c_A = B/(\mu_0\rho)^{\frac{1}{2}}$, and the waves below the ion cyclotron frequency as Alfvén waves. As might be inferred from their phase velocity, since it involves only the magnetic field and the electron plus ion fluid densities, it is essentially a hydromagnetic wave and can be likened to an elastic body wave with the magnetic field providing the energy storage analogous to the strain energy.

Above the ion cyclotron frequency and below the electron cyclotron frequency, only one branch propagates and the dispersion relation

$$\frac{c^2 k_z^2}{\omega^2} = 1 + \frac{\omega_{pe}^2 + \omega_{pi}^2}{(\omega + \omega_{ci})(\omega_{ce} - \omega)}$$

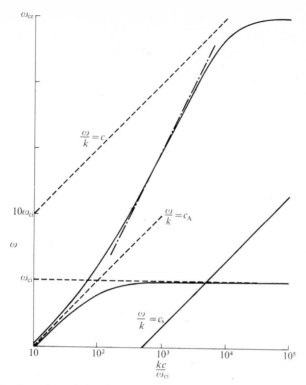

FIG. 9.1. The dispersion relations for low-frequency plasma waves propagating along a uniform magnetic field in a cold plasma. Near the cyclotron frequencies they are known as electron and ion cyclotron waves. At low frequencies two branches of oppositely rotating polarization merge to form the Alfvén wave with phase velocity c_A. The ion acoustic (longitudinal) wave is also shown for reference purposes. The dot–dash line of slope $\frac{1}{2}$ indicates the region where the whistler or helicon approximation $\omega = k^2(B/ne\mu_0)$ is appropriate. Parameters have been chosen so that $\omega_{ce} = 10^4\omega_{ci}$, $c_A = 10^{-2}c$, $c_s = 2\times10^{-2}c_A$.

can be approximated by

$$\frac{c^2k_z^2}{\omega^2} = \frac{\omega_{pe}^2}{\omega\omega_{ce}}$$

or

$$\omega = k_z^2\frac{B}{ne\mu_0}, \tag{9.4}$$

and this part of the dispersion curve has slope $\frac{1}{2}$, with the consequence that the phase velocity is higher for higher frequencies. This causes a descending audiofrequency signal observed in radiofrequency propagation in the ionosphere generated by lightning strokes, which has become known as a whistler.

Examining the nature of these waves in a little more detail, it is seen that since $\mathbf{E}$ is perpendicular to $\mathbf{k}$ they are transverse in the electrodynamic sense, and since $\mathbf{v}$ is perpendicular to $\mathbf{k}$ they are transverse in the fluid sense. While the phase velocities are significantly different for $\omega \to \omega_{ci}$ and therefore birefringence and the phenomenon of Faraday rotation can be expected, as $\omega \to 0$ the difference in phase velocities tends to zero, and thus the plane of polarization is preserved, i.e. a plane-polarized wave remains plane-polarized.

9.2. Propagation perpendicular to the magnetic field

The general equation for wave propagation in a medium with a relative permittivity ε is

$$\nabla \times \nabla \times \mathbf{E} = \frac{1}{c^2}\frac{\partial^2}{\partial t^2}(\varepsilon\mathbf{E}),$$

which with $\mathbf{E}$ of the form

$$(E_x\mathbf{i}+E_y\mathbf{j})\exp\{i(-\omega t + kx)\},$$

i.e. a plane wave propagating perpendicular to the magnetic field $B = B_0\mathbf{k}$, gives

$$\frac{c^2}{\omega^2}\{k^2\mathbf{E}-(\mathbf{k}.\mathbf{E})\mathbf{k}\} = \varepsilon\mathbf{E}.$$

In component form,

$$0 = \varepsilon_{11}E_x - i\varepsilon_{12}E_y,$$

$$\frac{k^2c^2}{\omega^2}E_y = i\varepsilon_{12}E_x + \varepsilon_{11}E_y,$$

which has as its solution

$$\frac{k^2c^2}{\omega^2} = \frac{(\varepsilon_{11}+\varepsilon_{12})(\varepsilon_{11}-\varepsilon_{12})}{\varepsilon_{11}}.$$

This can be written,

$$\frac{k^2c^2}{\omega^2} =$$

$$= \frac{(\omega^2-\omega_{ce}\omega_{ci}-\omega_p^2-\omega\omega_{ce}+\omega\omega_{ci})(\omega^2-\omega_{ce}\omega_{ci}-\omega_p^2+\omega\omega_{ce}-\omega\omega_{ci})}{(\omega^2-\omega_{ce}^2)(\omega^2-\omega_{ci}^2)-\omega_p^2(\omega^2-\omega_{ci}\omega_{ce})}, \quad (9.5)$$

and for $\omega \ll \omega_{ci}$ the right-hand side becomes

$$\frac{\omega_p^2+\omega_{ce}\omega_{ci}}{\omega_{ce}\omega_{ci}} = 1+\frac{\rho}{\varepsilon_0 B^2}.$$

Thus this wave has the same phase velocity as that of the wave propagating along B_0, but this wave has $\mathbf{k} \cdot \mathbf{E} \neq 0$ and $\mathbf{k} \cdot \mathbf{v}_e$, $\mathbf{k} \cdot \mathbf{v}_i \neq 0$ and hence is longitudinal in both senses; the wave is therefore described as the *compressional* or *fast Alfvén wave* and also as the *fast magnetoacoustic wave*.

Now let us examine the complete dispersion diagram. The numerator has zeros where $\omega^2 - \omega_p^2 - \omega\omega_{ce} \simeq 0$ and $\omega^2 - \omega_p^2 + \omega\omega_{ce} \simeq 0$ or

$$\omega^2 = \omega_p^2 \pm \omega_p \omega_{ce} \quad \text{for} \quad \omega_{ce} \ll \omega_{pe}$$

and

$$\omega = \frac{\omega_{ce}}{2}\left(1 + \frac{2\omega_p^2}{\omega_{ce}^2} \pm 1\right)$$

for $\omega_{ce} \gg \omega_{pe}$. The denominator has zeros for

$$\omega^2 \simeq \omega_p^2 + \omega_{ce}^2 \quad \text{and} \quad \omega^2 \simeq \frac{\omega_{ce}\omega_{ci}(\omega_p^2 + \omega_{ce}\omega_{ci})}{\omega_p^2 + \omega_{ce}^2},$$

and these are known as the upper and lower hybrid frequencies ω_{UH}, ω_{LH}. For $\omega_p \gg \omega_{ce}$, $\omega_{LH} \sim (\omega_{ce}\omega_{ci})^{\frac{1}{2}}$.

The low-frequency branch of the solution is shown in Fig. 9.2.

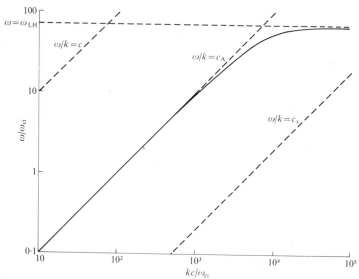

FIG. 9.2. The dispersion relation for the low-frequency plasma wave propagating exactly at right angles to a uniform magnetic field in a cold plasma, the compressional Alfvén wave. There is a resonance $k \to \infty$ at the lower hybrid frequency given by

$$\omega_{LH}^2 = \frac{\omega_{ce}\omega_{ci}(\omega_p^2 + \omega_{ce}\omega_{ci})}{\omega_p^2 + \omega_c^2},$$

where $\omega_p^2 = \omega_{pe}^2 + \omega_{pi}^2$, $\omega_c^2 = \omega_{ce}^2 + \omega_{ci}^2$. The parameters are the same as for Fig. 9.1.

9.3. Propagation at an angle to the magnetic field

In this case we proceed similarly but with $\mathbf{k} = k_z\hat{\mathbf{k}} + k_x\mathbf{i}$

$$\mathbf{k} = k_z\hat{\mathbf{k}} + k_x\mathbf{i} \equiv k\cos\theta\hat{\mathbf{k}} + k\sin\theta\mathbf{i}$$

and

$$\mathbf{E} = (E_x\mathbf{i} + E_y\mathbf{j} + E_z\mathbf{k})\exp\{i(-\omega t + k_z z + k_x x)\}.$$

The x, y, and z components of $k^2\mathbf{E} - (\mathbf{k}\cdot\mathbf{E})\mathbf{k} = (\omega^2/c^2)\boldsymbol{\varepsilon}\mathbf{E}$, become

$$k^2\cos^2\theta E_x - k^2\cos\theta\sin\theta E_z = \frac{\omega^2}{c^2}(\varepsilon_{11}E_x - i\varepsilon_{12}E_y),$$

$$k^2 E_y = \frac{\omega^2}{c^2}(i\varepsilon_{12}E_x + \varepsilon_{11}E_y),$$

$$-k^2 E_x\sin\theta\cos\theta + k^2 E_z = \frac{\omega^2}{c^2}\varepsilon_{33}E_z,$$

with the dispersion relation given by the determinant but which can be written for future comparison,

$$\left(k^2\cos^2\theta - \frac{\varepsilon_{11}\omega^2}{c^2}\right)\left(k^2 - \frac{\varepsilon_{11}\omega^2}{c^2}\right) - \varepsilon_{12}^2\frac{\omega^4}{c^4}$$

$$= k^4\sin^2\theta\cos^2\theta\,\frac{k^2 - \varepsilon_{11}(\omega^2/c^2)}{k^2\sin^2\theta - \varepsilon_{33}(\omega^2/c^2)}. \quad (9.6)$$

The low-frequency collisionless approximations to ε_{11} and ε_{12} are

$$1 + (c^2/c_A^2) \quad \text{and} \quad \omega\omega_{pe}^2/\omega_{ce}\omega_{ci}^2(\to 0).$$

So that

$$\frac{c^2 k^2}{\omega^2} \to \varepsilon_{11},\ \varepsilon_{11}\sec^2\theta \quad \text{or} \quad \frac{\omega^2}{k^2} = \frac{c_A^2 c^2}{c_A^2 + c^2}, \quad \frac{c_A^2 c^2}{c^2 + c_A^2}\cos^2\theta.$$

As $\cos\theta \to 0$ apparently only one wave remains, the faster wave of the two. We examine this branch further in the next section, noting that for large magnetic fields as the phase velocity is reduced it will become comparable to the ion sound speed $c_s = (k_B T_e/M)^{\frac{1}{2}}$.

The dispersion curve for $\theta = 45°$ is shown in Fig. 9.3.

A more convenient form for the dispersion equation (9.6), in general, is found by rearranging and solving for $\tan\theta$ to be

$$\tan^2\theta = -\frac{2\varepsilon_{33}\{(c^2 k^2/\omega^2) - \varepsilon_R\}\{(c^2 k^2/\omega^2) - \varepsilon_L\}}{\{(c^2 k^2/\omega^2) - \varepsilon_{33}\}\{(\varepsilon_R + \varepsilon_L)(c^2 k^2/\omega^2) - 2\varepsilon_R\varepsilon_L\}}. \quad (9.6a)$$

In this form the modes of propagation along the field lines, namely, the Langmuir oscillation and those discussed in § 9.1, appear as zeros of the

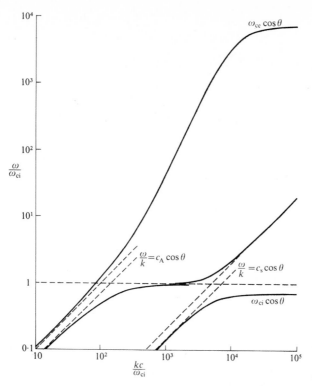

FIG. 9.3. Dispersion relations for low-frequency plasma waves propagating at an angle of 45° to a uniform magnetic field in a cold plasma. The fast Alfvén wave becomes, with increasing frequency, a slightly modified helicon wave and then a modified electron cyclotron wave. The slow Alfvén wave goes over at the ion cyclotron frequency into the electrostatic ion cyclotron wave (or first ion cyclotron wave), which in turn becomes an ion acoustic wave. There is also a branch entirely below the ion cyclotron frequency which is a modified ion acoustic wave and becomes a modified ion cyclotron wave (second ion cyclotron wave). Parameters as in Fig. 9.1. The two ion acoustic branches arise when the electron temperature is allowed to be finite.

numerator, and those perpendicular, namely the hybrid mode of § 9.2, together with a modified electromagnetic wave with **E** and **B** parallel, as zeros of the denominator.

9.4. Inclusion of electron temperature—ion cyclotron waves

In this section we examine further the low-phase velocity branch propagating at an angle to the magnetic field and specifically include electron temperature, i.e. electron pressure in the electron equation of motion. On the other hand, since $\omega_{ce} \gg \omega_{ci}$ we can neglect the effect of the magnetic field and electron inertia.

Thus the electron motion is given by

$$0 = e\mathbf{E} + \nabla(nk_B T_e),$$

and ion motion by

$$M\frac{d\mathbf{v}}{dt} = e(\mathbf{E} + \mathbf{v} \times \mathbf{B}_0).$$

Continuity requires

$$\frac{\partial n}{\partial t} + \nabla \cdot (n\mathbf{v}) = 0.$$

Substituting

$$n = n_0 + n_1 \exp\{i(-\omega t + k_x x + k_z z)\},$$

so that

$$\mathbf{E} = (E_x \mathbf{i} + E_z \mathbf{k})\exp\{i(-\omega t + k_x x + k_z z)\}$$

and

$$\mathbf{v} = (v_x \mathbf{i} + v_y \mathbf{j} + v_z \mathbf{k})\exp\{i(-\omega t + k_x x + k_z z)\};$$

gives

$$i\omega n_1 = in_0(k_x v_x + k_z v_z),$$

$$eE_x = -ik_B T_e k_x \frac{n_1}{n_0},$$

$$eE_z = -ik_B T_e k_z \frac{n_1}{n_0},$$

$$v_x = -eE_x/iM\omega\left(1 - \frac{\omega_{ci}^2}{\omega^2}\right), \qquad v_z = -\frac{eE_z}{iM\omega}$$

with the dispersion relation

$$\omega^2 = \frac{k_B T_e}{M}\left\{k_z^2 + k_x^2 \Big/ \left(1 - \frac{\omega_{ci}^2}{\omega^2}\right)\right\}, \tag{9.7}$$

or writing

$$k_x = k \sin\theta, \qquad k_z = k\cos\theta,$$

$$\omega^2 = \frac{k_B T_e}{M} k^2 \left(\frac{\omega^2 - \omega_{ci}^2 \cos^2\theta}{\omega^2 - \omega_{ci}^2}\right). \tag{9.8}$$

This has a resonance ($k \to \infty$) at $\omega = \omega_{ci} \cos\theta$, and a cut-off ($k \to 0$) at $\omega = \omega_{ci}$ with propagation below $\omega_{ci} \cos\theta$ and above ω_{ci}. The high-frequency branch becomes the ordinary ion acoustic wave. The low-frequency branch has a limiting velocity $c_s \cos\theta$.

Since the electric field of this wave is the gradient of a scalar quantity it is an electrostatic wave. The assumptions made in deriving it require $v_i \ll \omega_{ci}$ and $\omega \ll \omega_{pe}, \omega_{ce}$; for $\omega \gg \omega_{ci}$ it is essentially the ion sound wave with a transverse wavenumber, of equivalently, in finite geometry. As

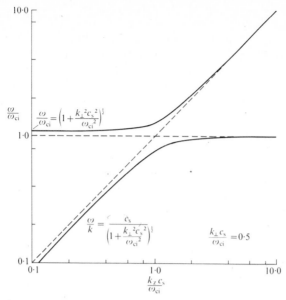

FIG. 9.4. Dispersion relations showing the coupling between ion cyclotron and ion acoustic modes in a uniform magnetic field corresponding to a fixed perpendicular wavenumber $k_\perp$ as the axial wavenumber k_z is varied. This represents an approximation to propagation on a finite plasma column taking $k_\perp \propto r_w^{-1}$ and shows a modified cut-off frequency ($k_z \to 0$) while the resonant frequency becomes ω_{ci} compared with $\omega_{ci} \cos \theta$ in Fig. 9.3.

$\omega \to \omega_{ci}$ from above the wavenumber goes to zero i.e. the phase velocity tends to infinity. In the case where $c_A > c_s$ the high-frequency branch joins onto the slow Alfvén wave and is known near ω_{ci} as the electrostatic ion cyclotron or first ion cyclotron wave. The low-frequency branch is often referred to as the slow magnetoacoustic wave below $\omega_{ci} \cos \theta$ and in the constant-frequency short-wavelength region as the second ion cyclotron wave.

The dispersion curves are shown in Fig. 9.4.

9.5. Experiments involving whistler and helicon waves

Measurements on the fast Alfvén branch of the dispersion relation have for a variety of reasons been carried out over different sections of the dispersion curve where different approximations are appropriate. The region below the ion cyclotron frequency, $\omega \lesssim \omega_{ci}$, will be left for discussion until § 9.6. In the range $3\omega_{ci} < \omega < 0.3\omega_{ce}$, as can be seen from Fig. 9.1, the dispersion relation is approximately

$$\omega = \frac{\omega_{ce} k^2 c^2}{\omega_{pe}^2} = \frac{k^2 B}{n e \mu_0},$$

and thus is determined by the number density and magnetic field, without inertia effects entering into the dispersion. Theoretical treatments based on the one-fluid model in this approximation have been given and experiments interpreted on this basis (see e.g. Klozenberg, McNamara, and Thonemann (1965) for a cylindrical plasma in vacuum, and Ferrari and Klozenberg (1968) for the case of a plasma bounded by conducting walls).

The name helicon has been applied to these waves in this frequency range, and since the dispersion is not significantly modified, even for the electron collision frequency $\nu_e \sim \omega$ these waves, amongst the whole range of plasma waves, are most readily observed in the solid state and have been extensively studied under such conditions (see Baynham and Boardman (1971) for a summary). Fig. 9.5 gives the measurements of Lehane and Thonemann (1965) compared with theory based on the one-fluid perfectly conducting plasma model.

There has been a steady relaxation of the assumptions made in the theory and therefore a more exact comparison with experiment since that time. Electron inertia was included by Davies (1969) and the effect of the radial variation of density through a scale length $L = (n \, \mathrm{d}r/\mathrm{d}n)$ by Davies and Christiansen (1969). Fig. 9.6 shows their theoretical results for the real and imaginary parts of the dispersion for the first-order $(m = 1)$ azimuthal mode in quite collisional plasmas, indicating that both plasma inhomogeneity and electron inertia have a measurable influence on the dispersion. Their measurements in a low-pressure mercury positive column are shown in Fig. 9.7, together with a comparison with theoretical curves for appropriate parameters. The conclusion is that electron inertia must be included to give a good description. Thus it seems more appropriate in treating these waves to begin with the two-fluid model and then make approximations at a later stage. Such an analysis is given in Appendix III (p. 229) and, amongst other results, reproduces by a slightly more elegant method results obtained by McKay (1967). McKay was able to identify the nature of the waves involved in the finite cylindrical situation as being modified infinite plasma helicon waves coupled with quasistatic modes of the Trivelpiece and Gould type (§ 6.4) propagating at low frequencies. He was also able to explain the apparent anomaly noted by Klozenberg, McNamara, and Thonemann that the damping was not a strong function of the collision frequency by carrying the analysis through in the range where $\nu_e < \omega$. Fig. 9.8 shows that the damping is approximately proportional to ν_e/ω while $\nu_e < \omega$ and then constant until ν_e becomes comparable with ω_{ce} when it again increases. The region $\nu_e < \omega$ was not accessible in the earlier work because of the initial approximations.

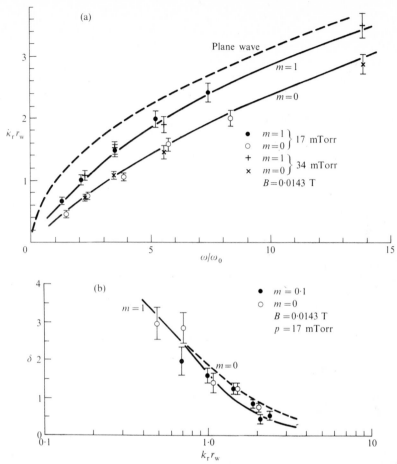

FIG. 9.5. Comparison between experiment and theory for helicon waves in a bounded low-pressure argon plasma column: (a) real part of the dispersion; (b) damping in terms of the decrement δ. The solution for an infinite plasma is shown dashed in (a). Both the symmetric ($m = 0$) and dipolar ($m = 1$) modes were excited and comparison is made at two pressures showing no significant influence in (a). The theoretical damping curves (b) are shown for the best-fit parameter $\omega_{ci}/\nu_i = 30$.

9.6. Electron cyclotron waves

For wave frequencies above about 0.3 of the electron cyclotron frequency the dispersion is principally determined by the cold-plasma resonance at $\omega = \omega_{ce}$. A certain amount of confusion has been caused by the use of the term 'whistler wave' in connection with this part of the dispersion curve when the ionospheric observations relate to the frequency region described in the previous section. As can be seen from Fig. 9.1, in this region the phase velocity (a local line of slope 1 on a

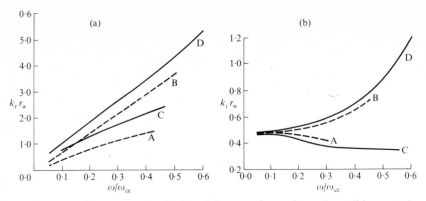

FIG. 9.6. Theoretical curves showing the influence of experimental conditions on the dispersion and spatial damping of helicon waves. Curve A is for a uniform plasma neglecting electron inertia $(m \to 0)$; curve B includes the effect of electron inertia taking $\omega_{ce} r_w / c_A = 12 \cdot 8$; curve C shows the effect of a radial variation of density by taking $n = n_0 J_0 (1 \cdot 7 r / r_w)$ for $\omega_{ce} / \omega \to \infty$; curve D includes the effects of non-uniformity as in C and electron inertia as in B. The finite value of k_i as $\omega \to 0$ is discussed later in this section.

logarithmic plot) is a *decreasing* function of frequency, and so a wave packet would have a rising tone. Laboratory experimental measurements demonstrating propagation of such waves were carried out at microwave frequencies on high-current pulsed discharges by Gallet, Richardson, Wieder, Ward, and Harding (1960) and Dellis and Weaver (1963), for instance, but it is only relatively recently that measurements of the dispersion have been made.

Lee, Fessenden, and Crawford (1969) using the afterglow of a pulsed cylindrical discharge at 1–5 mTorr in argon, obtained real and imaginary

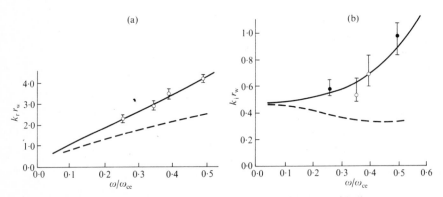

FIG. 9.7. Experimental measurements by Davies and Christiansen of helicon waves on a mercury column $p = 2$ mTorr, $r_w = 35$ mm compared with theory neglecting electron inertia (dashed line) and with the value of ω_{ce} / ω appropriate to a magnetic field $0 \cdot 002$ Torr (full line), the density was taken to be $4 \cdot 8 \times 10^{17}$ m^{-3} and $\omega_{ce} / \nu_e = 6 \cdot 8$.

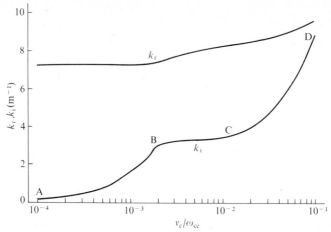

FIG. 9.8. The effect of collisions on the real and imaginary parts of the dispersion of helicon waves for fixed frequency $\omega/2\pi = 9\cdot48 \times 10^6$ Hz as computed by McKay for a uniform plasma of radius 0·05 m with $\omega_{pe}/2\pi = 10^{11}$ Hz and $\omega_{ce}/2\pi = 2\cdot02 \times 10^9$ Hz in a waveguide showing that for $\nu_e \ll \omega$ the damping tends to zero and that as $\nu_e \to \omega_{ce}$ both real and imaginary parts are affected while in the portion BC there is a region where the damping is independent of ν_e/ω_{ce} and this corresponds to the finite k_i as $\omega \to 0$ in Figs 9.6(b) and 9.7(b).

wavenumber as a function of wave frequency and were able to show the increase in damping expected as the cyclotron frequency is approached.

A kinetic-theory treatment for an infinite plasma with a Maxwellian velocity distribution gives a dispersion relation

$$\frac{k^2c^2}{\omega^2} = 1 + \frac{\omega_{pe}^2}{\omega k c_e} Z\left(\frac{\omega - \omega_{ce} + i\nu_e}{k c_e}\right), \qquad (9.9)$$

where Z is the plasma dispersion function. Thus there is damping analogous to Landau damping for $\omega - \omega_{ce} \sim k c_e$, i.e. when there is resonance between the phase velocity corresponding to the wave frequency Doppler shifted into the electron frame, and the electron thermal speed. This damping is usually referred to as cyclotron damping, and the expected form of the dispersion relation near the electron cyclotron frequency when electron temperature is included is shown in Fig. 9.9 for a relatively dense plasma. Corresponding experimental results are given in Fig. 9.10 but it was not possible to distinguish between collisional and cyclotron damping. However, this has been achieved recently by Christopoulos, Boswell, and Christiansen (1974) in a radiofrequency-maintained discharge.

Ion cyclotron waves above the ion cyclotron frequency do not appear to have been studied in active discharges but their general features of Fig. 9.4 have been confirmed recently in Q-machine plasmas (Sato, Sugai,

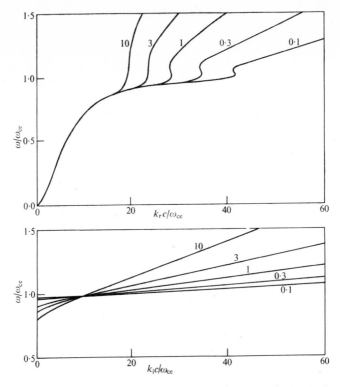

FIG. 9.9. Theoretical dispersion and spatial damping curves for electron cyclotron waves including the effects of finite electron temperature, with values ranging from 0·1 eV to 10 eV while $\omega_{pe}^2/\omega_{ce}^2$ has the fixed value 40. The upper branch of Fig. 9.1 is modified by interaction with the electrons characterized by their thermal speed c_e. Heavy damping sets in close to and above the electron cyclotron frequency ω_{ce}.

Sasaki, and Hatakeyama 1974; Edgley, Franklin, Hamberger, and Motley 1975). Experiments to study propagation below the ion cyclotron frequency will be considered in the next section.

9.7. Experimental measurement of Alfvén waves

Before discussing the very full range of measurements that have been made on this low-frequency part of the dispersion curve, it is necessary to introduce an important additional consideration. This is necessary because the measurements have been made on dense partially ionized plasmas, and since ion motion is involved there can be a significant transfer of momentum from the ions to the neutral gas. This is commonly referred to as gas-loading and means that the treatment has to be effectively a three-fluid one, with distinction being made between the different types of collisional encounter, so that the equations of motion

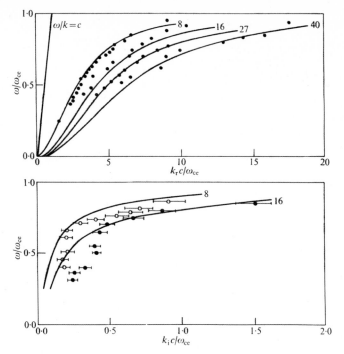

FIG. 9.10. Measured dispersion and damping of electron cyclotron waves in argon compared with theory for different values of the parameter ω_{pe}^2. Plasma conditions: 1 mTorr, $T_e = 0 \cdot 1$ eV, $\omega_{\text{ce}}/2\pi = 4$ GHz.

become

$$n_e m_e \frac{\partial \mathbf{v}_e}{\partial t} = -n_e e(\mathbf{E} + \mathbf{v}_e \times \mathbf{B}) + m_e n_e \nu_{\text{ei}}(\mathbf{v}_i - \mathbf{v}_e), \tag{9.10}$$

$$n_i m_i \frac{\partial \mathbf{v}_i}{\partial t} = n_i e(\mathbf{E} + \mathbf{v}_i \times \mathbf{B}) + m_i n_i \nu_{\text{in}}(\mathbf{v}_n - \mathbf{v}_i) + m_i n_i \nu_{\text{ie}}(\mathbf{v}_e - \mathbf{v}_i), \tag{9.11}$$

$$n_n m_n \frac{\partial \mathbf{v}_n}{\partial t} = m_n n_n \nu_{\text{ni}}(\mathbf{v}_i - \mathbf{v}_n), \tag{9.12}$$

where the subscript n refers to neutral atoms.
    For motion with a frequency ω eqn (9.12) becomes

$$\mathbf{v}_i = \mathbf{v}_n\left(1 - \frac{i\omega}{\nu_{\text{ni}}}\right),$$

and the second term on the right in (9.11) becomes

$$-m_i n_i \nu_{\text{in}} \mathbf{v}_i \frac{i\omega}{-\nu_{\text{ni}} + i\omega}, \quad \text{and since} \quad \frac{\nu_{\text{ni}}}{n_i} = \frac{\nu_{\text{in}}}{n_n}$$

the ion equation of motion becomes

$$m_i n_i \mathbf{v}_i \left(-i\omega + \frac{i\omega \nu_{ni}}{-\nu_{ni}+i\omega} \frac{n_n}{n_i} \right) = n_i e (\mathbf{E} + \mathbf{v}_i \times \mathbf{B}) + \mathbf{P}_{ie},$$

where $\mathbf{P}_{ie}$ is the momentum transfer from ions to electrons and $\mathbf{P}_{ie} = -\mathbf{P}_{ei}$. This modifies the inertia and effective collision frequency of the ions. So that (9.6) becomes, on writing $k_z = k \cos \theta$, $k_\perp = k \sin \theta$ and taking the low-frequency ($\omega \sim \omega_{ci}$) and low-phase-velocity ($\omega/k \ll c$) approximation,

$$\left(S k_z^2 - \frac{\omega^2}{c_A^2} \right) \left\{ S(k_z^2 + k_\perp^2) - \frac{\omega^2}{c_A^2} \right\} = k_z^2 (k_z^2 + k_\perp^2) \frac{\omega^2}{\omega_{ci}^2}, \qquad (9.13)$$

where

$$S = \frac{n_i + i n_n (\omega/\nu_{in})}{n_i + n_n + i n_n (\omega/\nu_{in})}.$$

Woods (1962) has given by far the most complete treatment of low-frequency waves in a cylindrical plasma and included, in addition to what is explicit or implicit in the equations above, the effects of particle temperatures and viscosity. There is also a careful discussion of the boundary conditions to be applied, which essentially determine the relation between $k_\perp$ and the plasma radius a. For finite conductivity with the conductivity tensor written

$$\begin{bmatrix} \sigma_\perp & 0 & 0 \\ 0 & \sigma_\perp & 0 \\ 0 & 0 & \sigma_\parallel \end{bmatrix}$$

the dispersion equation becomes

$$\left\{ S k_z^2 - \frac{\omega^2}{c_A^2} \left(1 + \frac{i k_z^2}{\sigma_\perp \omega} + \frac{i k_\perp^2}{\sigma_\parallel \omega} \right) \right\} \left\{ S(k_z^2 + k_\perp^2) - \frac{\omega^2}{c_A^2} \left(1 + \frac{i k_z^2}{\sigma_\perp \omega} + \frac{i k_\perp^2}{\sigma_\parallel \omega} \right) \right\}$$

$$= k_z^2 (k_z^2 + k_\perp^2) \frac{\omega^2}{\omega_{ci}^2}. \qquad (9.14)$$

The physical conditions required for measurements on Alfvén waves are that both electrons and ions should be magnetized or $r_w > v_i/\omega_{ci}$, implying $B > 0.5$ Torr and that the wave phase velocity should be low enough so that wavelengths do not require an apparatus of a physically impossible length. Suppose that $k \sim 10 \, \mathrm{m}^{-1}$, then the charged-particle densities required are $\sim 10^{22} \, \mathrm{m}^{-3}$. This is outside the range of normal active discharges and therefore measurements have been possible only in pulsed discharges, but have then been made while current is still flowing and in the afterglow. Excitation and detection are carried out in terms of perturbations to the magnetic field. Fig. 9.11 shows results obtained by

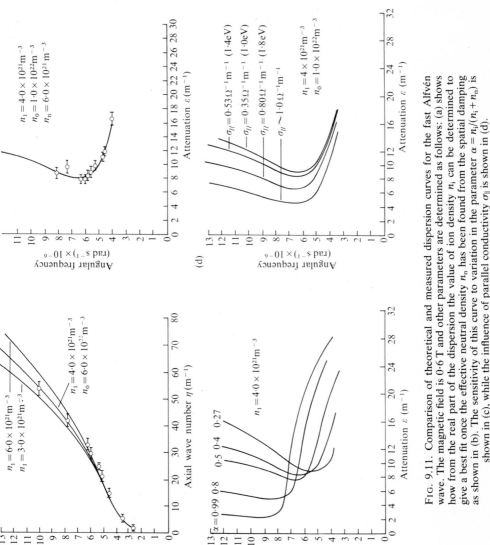

Fig. 9.11. Comparison of theoretical and measured dispersion curves for the fast Alfvén wave. The magnetic field is 0·6 T and other parameters are determined as follows: (a) shows how from the real part of the dispersion the value of ion density n_i can be determined to give a best fit once the effective neutral density n_n has been found from the spatial damping as shown in (b). The sensitivity of this curve to variation in the parameter $\alpha = n_i/(n_i + n_n)$ is shown in (c), while the influence of parallel conductivity $\sigma_{\parallel}$ is shown in (d).

Malein (1965) in hydrogen compared with theory showing best-fit results for both the real and imaginary parts of the wavenumber and indicating the sensitivity of the dependence on ion density, neutral density, and conductivity. Earlier measurements of the fast Alfvén wave were reported by Swanson, Gould, and Hertel (1964).

Subsequent work has included the effect of a radial distribution of the plasma density when the radial perturbation to the magnetic field B_{1r} satisfies

$$\frac{\partial^2 B_{1r}}{\partial r^2} + \frac{1}{r}\frac{\partial B_{1r}}{\partial r} + \left(k_\perp^2 - \frac{1}{r^2}\right)B_{1r} = 0,$$

and boundary condition $B_{1r} = 0$ at the conducting wall determines $k_\perp$ in terms of roots of Bessel functions. Solving (9.6) for $k_\perp^2$ gives

$$\frac{\omega^2 \varepsilon_{11}}{c^2} - k_z^2 + \frac{k_z^2(k_z^2 + k_\perp^2)(\omega^2/\omega_{ci}^2)}{k_z^2 - (\omega^2/c^2)\varepsilon_{11}},$$

and treating ε as a function of r the differential equation can be solved to find the modified eigenvalue. For infinite conductivity, zero gas-loading, and a parabolic density variation $n_e = n_0(1 - (r^2/r_w^2))$ the solution can be expressed in terms of confluent hypergeometric functions, but otherwise one has to resort to numerical methods. Such an analysis of experimental results has been given by Morrow and Brennan (1971).

9.8. The slow Alfvén wave

The dispersion relations given in the preceding section also describe the slow Alfvén wave which has a resonance as $\omega \to \omega_{ci}$ for a cold plasma. This mode was discovered earlier experimentally (Allen, Baker, Pyle, and Wilcox 1959) and also has been extensively measured. The comparison obtained between experiment (Jephcott and Stocker 1962) and theory (Woods 1962) for helium and argon at frequencies up to ω_{ci} is shown in Fig. 9.12. The transition into the electrostatic ion cyclotron wave does not appear to have been demonstrated.

9.9. Non-axisymmetric Alfvén waves

A further contribution to the detailed understanding of Alfvén waves has come from studies of the propagation of higher-order modes, i.e. those of the form $B_{1z} \propto J_m(k_\perp r)\exp\{i(-\omega t + kz + m\theta)\}$ with $m \neq 0$ and, in particular, $m = \pm 1$. The sensitivity of the dispersion and damping to the radial density distribution has been examined by Lehane and Paoloni (1972). Fig. 9.13 shows measured phase velocity and damping for the $m = 0$ modes and the $m = +1$ and -1 modes compared with theoretical results for two cases: (1) uniform plasma filling the metal tube which

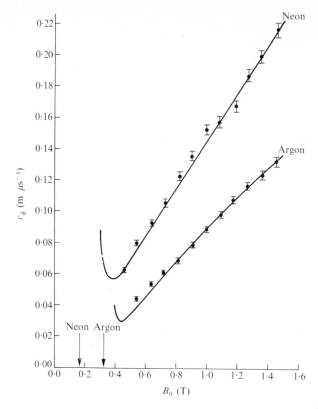

FIG. 9.12. A comparison of experiment and theory for the slow Alfvén wave in neon and argon with parameters n_i, n_n, ν_i, $\sigma_\parallel$ chosen to give best fit. In this case the phase velocity $v_\phi = \omega/k$ is plotted as a function of the magnetic field. Resonance at $\omega = \omega_{ci}$ can be seen from Fig. 9.1 to imply a vanishing phase velocity, and the corresponding magnetic fields for $\omega/2\pi = 125$ kHz are indicated by arrows.

enclosed the glass vacuum vessel; (2) a cold gas layer of variable thickness t (treated as being of relative permittivity unity, i.e. as if it were vacuum). Clearly a better description can be obtained for a specific value of t.

9.10. Ion Bernstein waves

In Chapter 6 it was shown that a kinetic treatment of the electron motion in a magnetized plasma gives rise to modes which propagate near the electron cyclotron harmonics and which are heavily damped except for propagation vectors almost at right angles to the magnetic field. Similar waves will exist under appropriate conditions for ions and the

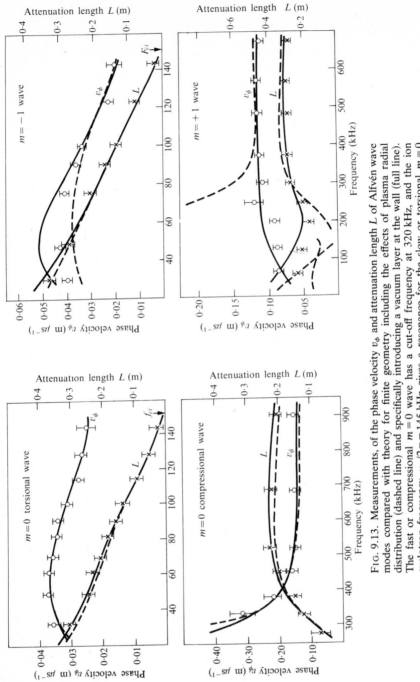

FIG. 9.13. Measurements, of the phase velocity v_ϕ and attenuation length L of Alfvén wave modes compared with theory for finite geometry including the effects of plasma radial distribution (dashed line) and specifically introducing a vacuum layer at the wall (full line). The fast or compressional $m = 0$ wave has a cut-off frequency at 320 kHz, and the ion cyclotron frequency $\omega_{ci}/2\pi = 145$ kHz gives a resonance for the slow or torsional $m = 0$ wave and $m = -1$ wave at that frequency.

generalization of eqn (9.6) can readily be made (see e.g. Stix 1962) to

$$\varepsilon(\omega, k_z, k_\perp) = 1 + \sum_{\alpha=e,i} \frac{1}{k^2 \lambda_{D\alpha}^2} \exp(-\Lambda_\alpha) \sum_{-\infty}^{\infty} I_n(\Lambda_\alpha)\{1 + \zeta_{0\alpha} Z(\zeta_{n\alpha})\},$$

where

$$k^2 = k_z^2 + k_\perp^2, \qquad \Lambda_\alpha = k_\perp^2 (k_B T_\alpha / m_\alpha \omega_{c\alpha}^2), \qquad \zeta_{n\alpha} = (\omega + n\omega_{c\alpha})/k_z c_\alpha.$$

Simplifications appropriate to ion waves propagating nearly perpendicular to the field are $k_z \rho_i \ll 1$, $\omega \sim n\omega_{ci}$, $k_\perp \rho_e \ll 1$, and these yield

$$\varepsilon(\omega, k_z, k_\perp) = 1 - \frac{1}{2k^2\lambda_D^2} Z'\left(\frac{\omega}{k_z c_e}\right) - \frac{1}{k^2\lambda_D^2} \exp(-\Lambda_i) \sum_{-\infty}^{\infty} I_n(\Lambda_i) \frac{n\omega_{ci}}{\omega - n\omega_{ci}}.$$

The contribution of the electrons then depends on the axial phase velocity and thus changes significantly as k_z passes through ω/c_e. For small k_z,

$$Z' \sim k_z^2 c_e^2 / \omega^2.$$

and the mode is described as the pure ion Bernstein wave while, for large

$$k_z \gg \omega \left(\frac{m}{2k_B T_i}\right)^{\frac{1}{2}},$$

the mode is called the neutralized ion Bernstein wave, since in this case the electrons are in equilibrium with the potential of the wave. In between there is significant Landau damping on the electrons due to the high axial phase velocity.

Fig. 9.14 gives comparative dispersion curves for both types of wave for specific conditions, where it can be seen that there is a characteristic ion hybrid frequency $\omega_{Hi}^2 = \omega_{pi}^2 + \omega_{ci}^2$ for the pure ion Bernstein waves and, as with the electron Bernstein waves (Fig. 6.14, p. 133), the form of the branches of the dispersion curve changes as the hybrid frequency is exceeded.

The most complete measurements of these waves have been made by Schmitt (1972, 1973a, 1973b) in a Q-machine plasma, demonstrating the features of the curves of Fig. 9.14 for both modes, including the influence of isotopes. Earlier measurements of Ault and Ikezi (1970) in microwave-generated low-pressure helium and argon plasmas demonstrated the existence of the neutralized ion Bernstein mode; their results are shown in Fig. 9.15.

Landau damping of ion Bernstein waves at the intermediate range of phase velocities (or angles of propagation) discussed above has been measured by Hirose, Alexeff, and Jones (1970), exploiting the arrangement and techniques described in connection with ion waves in Chapter 7. They were particularly concerned to show the effects of mixtures of ions on the propagation of ion cyclotron modes and that the presence of a

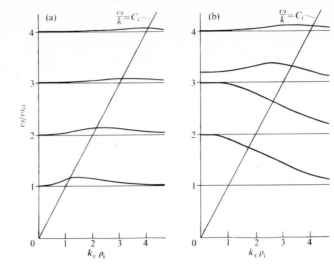

FIG. 9.14. Theoretical dispersion curves for ion Bernstein waves showing the differences between pure ion Bernstein waves and neutralized ion Bernstein waves. The parameters used are $\omega_{pi}/\omega_{ci} = 3$ and $T_e/T_i = 1$, and $\omega_{hi}^2 \equiv \omega_{pi}^2 + \omega_{ci}^2$ is the ion hybrid frequency.

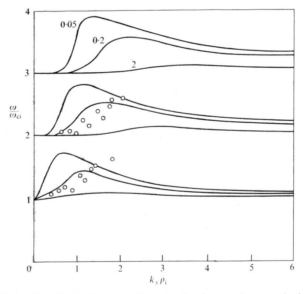

FIG. 9.15. Measured ion Bernstein-wave dispersion for frequencies near the ion cyclotron frequency ω_{ci} and its second harmonic compared with theory for the neutralized modes with T_i/T_e as a parameter. The plasma was a microwave discharge $r_w = 15$ mm in helium at $\sim 0 \cdot 1$ mTorr.

light contaminant ion can cause enhanced damping of modes involving a heavier ion.

9.11. Concluding remarks

The general conclusion to be drawn from the measurements reported in this chapter is that when the effects of resistivity, collisions, neutral gas inertia, and radial inhomogeneity are taken into account, a good description can be given of axially propagating plasma waves at low frequencies in bounded plasmas. There is one further factor which is of importance under some circumstances in inhomogeneous plasmas in a magnetic field and this arises from the particle drifts described in Chapter 3 and which are responsible for the diamagnetism of a plasma. Drift waves are discussed in Chapter 10.

10

DRIFT WAVES

10.1. Drift waves as modified ion waves

WE HAVE seen in Chapter 3 that the presence of radial number density gradients and radial electric fields leads to azimuthal motion of the charged particles in the presence of a magnetic field. This azimuthal motion can be expected to influence the propagation of waves which have a component of their propagation vector in the direction of this drift, and also have phase velocities comparable to it.

Thus we take the equations which led to ion waves in Chapter 7 and modify them by including a steady electron drift. Taking, as before, the z-axis to be the direction of the magnetic field and supposing a density gradient in the x-direction leads to an electron velocity v_{0y} in the y-direction given by

$$-ev_{0y}B_z = \frac{k_B T_e}{n_0}\frac{dn_0}{dx}. \tag{10.1}$$

For the perturbed variables associated with the wave, we use the electrostatic approximation to relate the electric field to a potential

$$\phi = \phi_1 \exp\{i(k_z z + k_y y - \omega t)\},$$

since the wave phase velocity is small compared with c, and take the perturbed densities equal and given by

$$n = n_1 \exp\{i(k_z z + k_y y - \omega t)\}$$

and the perturbed velocity

$$\mathbf{v} = \mathbf{v}_1 \exp\{i(k_z z + k_y y - \omega t)\}.$$

The continuity equation becomes

$$-i\omega n_1 + n_0 i(v_{1z}k_z + v_{1y}k_y) + v_{1x}\frac{dn_0}{dx} = Zn_1;$$

using $v_{i0} = 0$ and $\nabla \cdot v_{i0} = 0$, the ion momentum is given by

$$i\omega M v_{1y} = iek_y\phi_1 + ev_{1x}B, \qquad i\omega M v_{1z} = iek_z\phi_1, \qquad -i\omega M v_{1x} = ev_{1y}B,$$

and the electron motion integrates to

$$\frac{n_1}{n_0} = \frac{e\phi_1}{k_B T_e}.$$

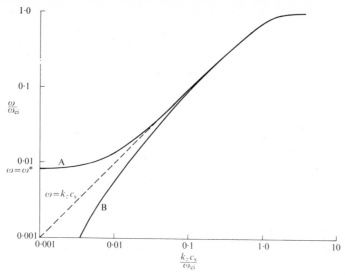

FIG. 10.1. The theoretical dispersion relations for low-frequency waves propagating at an angle to the magnetic field in an inhomogeneous magnetized plasma showing how the dispersion is significantly modified for the branch for which the electron drift is in the direction of the wave. Frequency is normalized to the ion cyclotron frequency ω_{ci} and the wavenumber to the ion sound speed c_s divided by ω_{ci}. The parameters $k_y v_0 \equiv k_y \kappa_n k_B T_e/B$ and $k_y c_s/\omega_{ci}$ are taken to be $0 \cdot 01 \omega_{ci}$ and $0 \cdot 5$ respectively.

Setting $\omega \ll \omega_{ci}$ these can readily be combined to give the dispersion equation

$$\omega(\omega - iZ - v_{0y} k_y) = k_z^2 \frac{k_B T_e}{M_i}. \tag{10.2}$$

The approximation made in setting the perturbed densities equal can readily be shown, as with ion waves, to restrict the validity of the treatment to values of $k\lambda_D < 1$. Adding the effect of collisions on the ion motion replaces the first ω by $\omega + i\nu_i$ and, in a situation where the plasma is not maintained by ionization by electron impact, Z should be set to zero.

Setting $Z = 0$ but allowing $\omega \sim \omega_{ci}$ gives the relation

$$\omega(\omega - k_y v_{0y}) = c_s^2 \left(k_z^2 + \frac{\omega^2 k_y^2}{\omega^2 - \omega_{ci}^2} \right), \tag{10.3}$$

and this is shown in Fig. 10.1. As $k_z \to 0$ the two branches are

$$\omega \to 0, \qquad k_y v_{0y} \Big/ \left(1 + \frac{k_y^2 c_s^2}{\omega_{ci}^2} \right)$$

and for large k_z, $\omega \to \pm k_z c_s$. The modification to the ion wave dispersion is thus seen to be the introduction of a cut-off frequency ω^* for waves propagating in the direction of v_0; the oppositely propagating wave is little affected by the existence of the azimuthal drift.

In order to make comparison with experimental measurements on plasma columns, it is necessary to take into account two additional features. The first is the existence of an electric field in the radial direction which causes an $|\mathbf{E} \times \mathbf{B}|/B^2 = v_\theta$ drift of particles of both signs. If E_r were proportional to r this would result in a rigid-body rotation of the column and this approximation is often employed in order to simplify the problem and avoid the need to treat all quantities as functions of the radius. The second is the problem of assigning an appropriate value to the quantity k_y. In a complete treatment in cylindrical coordinates the perturbed quantities would be taken to vary as $\exp im\theta$ and single-valuedness would be achieved by taking m to be an integer. A rough approximation can be made by assigning a radius r_0 and requiring $k_y r_0$ to be an integer. The most appropriate value would seem to be the radius of maximum wave amplitude, and the dispersion can be made to appear not to depend on r_0 by introducing the rotation frequency at r_0 defined by

$$\omega_r = v_0/r_0 = mk_y v_0. \quad .$$

The continuity equation is now

$$\frac{\partial n_1}{\partial t} + n_1 \frac{dv_0}{dx} + n_0 \mathbf{ik} \cdot \mathbf{v}_1 + v_{0y} \frac{dn_1}{dy} + v_{1x} \frac{dn_0}{dx} = 0$$

and the ion motion is given by

$$(-i\omega + \nu_i + ik_y v_{0y})v_{1x} = \frac{e}{M} v_{1y} B,$$

$$(-i\omega + \nu_i + ik_y v_{0y})v_{1y} = -\frac{e}{M} v_{1x} B - \frac{e}{M} ik_y \phi_1,$$

$$(-i\omega + \nu_i + ik_y v_{0y})v_{1z} = -\frac{e}{M} ik_z \phi_1,$$

while the steady-state ion motion equations give $\mathbf{v}_0$ in terms of the steady-state radial field alone, since

$$v_{0x}\nu_i = -\frac{e}{M} \frac{d\phi_0}{dx} + \frac{e}{M} v_{0y} B \quad \text{and} \quad v_{0y}\nu_i = -\frac{e}{M} v_{0x} B$$

yield

$$v_{0y} = \frac{\omega_{ci}^2}{\nu_i^2 + \omega_{ci}^2} \frac{1}{B} \frac{d\phi_0}{dx}.$$

In the rigid-body approximation $dv_0/dx = 0$ and the continuity equation yields

$$(\omega - \omega_r)\frac{n_1}{n_0} = \frac{e\phi_1}{M}\left(\frac{k_z^2}{\omega_1} + \frac{k_y^2\omega_1 + k_y\kappa_n\omega_{ci}}{\omega_1^2 - \omega_{ci}^2}\right),$$

with the notation $\omega_1 = \omega - \omega_r + i\nu_i$ and $\kappa_n = (1/n_0)(dn_0/dx)$.

The dispersion relation in the absence of collisions is then (with $\omega' = \omega - \omega_r$)

$$c_s^2\left\{\frac{k_z^2}{\omega'^2} + \frac{k_y^2 + k_y\kappa_n(\omega_{ci}/\omega')}{\omega'^2 - \omega_{ci}^2}\right\} = 1, \qquad (10.4)$$

which, writing

$$\omega^* = |\kappa_n|\frac{k_y k_B T_e}{eB},$$

can be seen to have $k_z = 0$ solutions of $\omega = \omega_r$ and $\omega \approx \omega_r + \omega^*$ and for large k_z given by $\omega \approx \pm k_z c_s + \omega_r + (\omega^*/2)$, when the approximations $\omega \ll \omega_{ci}$ and $k_y \ll (\omega_{ci}/c_s)$ are used.

The most comprehensive comparison with experiment has been made by Keen and Aldridge (1970), who measured the electric field, the number density and its gradient, and the rotation frequency as well as the propagation characteristics of low-frequency waves. Fig. 10.2 shows their results compared with theory for drift-waves propagating with and against the magnetic field and also for $k_y = 0$, i.e. the ion wave.

10.2. Drift waves at high phase velocities

We have seen in the case of homogeneous plasmas that the dispersion is modified whenever the phase velocity of a branch becomes of the order of one of the characteristic speeds either of particles c_e and c_i, or of wave modes c_s, c_A, and c. Thus it is to be expected that the upper branch of Fig. 10.1 will be modified as $\omega/k \to c_A$ for magnetic fields sufficiently large that $c_A > c_s$. The electrostatic approximation is no longer valid or appropriate and therefore the full Maxwell's equations must be used. Furthermore, to ease the manipulation we will treat the plasma as collisionless. Otherwise the model is as before, with a steady-state electron velocity v_{0y} given by

$$v_{0y} = -\frac{k_B T_e}{B_0}\frac{1}{n_0}\frac{dn_0}{dx} = -\kappa_n\frac{k_B T_e}{B_0}.$$

Writing the number density

$$n = n_0 + n_1 \exp\{i(-\omega t + k_z z + k_y y)\},$$

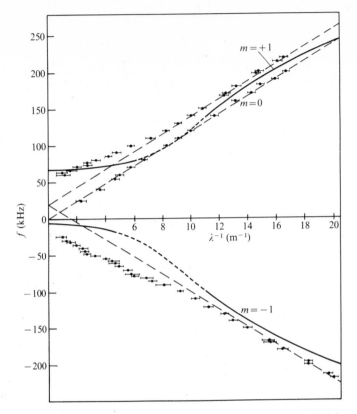

F$_{IG}$. 10.2. Dispersion of drift waves measured by Keen and Aldridge (1970) in a helium plasma column at 0·25 mTorr, axial field 0·1 T, showing a comparison between theory and experiment for the two drift modes and also for the mode with $k_y = 0$, the normal ion wave. The dashed curves are the simple analytical approximations.

assuming quasi-neutrality from the outset, and setting

$$\mathbf{E}_1 = (E_y\mathbf{j} + E_z\mathbf{k})\exp\{i(-\omega t + k_z z + k_y y)\},$$

$$\mathbf{v}_e = v_{0y}\mathbf{j} + (v_{e1x}\mathbf{i} + v_{e1y}\mathbf{j} + v_{e1z}\mathbf{k})\exp\{i(-\omega t + k_z z + k_y y)\}.$$

$\mathbf{B}_1$ is given by

$$\frac{\partial \mathbf{B}_1}{\partial t} = -\nabla \times \mathbf{E}_1,$$

and thus

$$\mathbf{B} = B_0\mathbf{k} + \frac{(k_y E_z - k_z E_y)}{\omega}\mathbf{i}.$$

The electron equation of motion neglecting electron inertia is

$$0 = -ne(\mathbf{E} + \mathbf{v} \times \mathbf{B}) - \nabla(nk_B T_e),$$

which from its z-component yields

$$\frac{i n_1}{n_0} = -\frac{e}{k_z k_B T_e}\left\{ E_z - \frac{v_{0y}}{\omega}(k_y E_z - k_z E_y)\right\}. \tag{10.5}$$

The ion equation of motion

$$M\frac{\partial \mathbf{v}_i}{\partial t} = e(\mathbf{E} + \mathbf{v}_i \times \mathbf{B}),$$

together with the ion equation of continuity

$$\frac{\partial n}{\partial t} + \nabla \cdot (n\mathbf{v}_i) = 0,$$

$\Bigg($ which in component form is

$$-i\omega M v_{ix} = e v_{iy} B_0, \qquad -i\omega M v_{iy} = e(E_y - v_{ix}B_0),$$

$$-i\omega M v_{iz} = eE_z, \qquad -i\omega n_1 + v_{ix}\frac{dn_0}{dx} + i n_0(k_y v_{iy} + k_z v_{iz}) = 0\Bigg)$$

gives

$$\frac{i n_1}{n_0} = -\frac{k_z e E_z}{M\omega^2} - \frac{k_y e E_y}{M(\omega^2 - \omega_{ci}^2)} - \frac{\kappa_n \omega_{ci} e E_y}{M\omega(\omega_{ci}^2 - \omega^2)}. \tag{10.6}$$

Eliminating n_1/n_0 between eqns (10.5) and (10.6) gives, for $\omega_{ci} \gg \omega$,

$$\frac{E_z}{k_z k_B T_e}\left(1 - \frac{v_0 k_y}{\omega} - \frac{k_z^2 c_s^2}{\omega^2}\right) = -\frac{k_y E_y}{M\omega_{ci}^2}. \tag{10.7}$$

A further relation between the electric field components can be found from the Maxwell relation

$$\nabla \times \mathbf{H} = \varepsilon_0 \frac{\partial \mathbf{E}}{\partial t} + \mathbf{J}.$$

Ignoring the displacement current and taking $\mathbf{J} = n_0 e \mathbf{v}_{il}$, as we know is appropriate for Alfvén waves, the y-component gives

$$k_z^2 E_y - k_y k_z E_z = \frac{\omega^2 \mu_0 n_0 e^2 E_y}{\omega_{ce}^2 M}$$

or

$$\left(k_z^2 - \frac{\omega^2}{c_A^2}\right)E_y = k_z k_y E_z. \tag{10.8}$$

Eqns (10.7) and (10.8) thus give a dispersion relation

$$(\omega^2 - k_z^2 c_A^2)\left(1 - \frac{k_y v_0}{\omega} - \frac{k_z^2 c_s^2}{\omega^2}\right) = \frac{c_s^2 c_A^2 k_y^2 k_z^2}{\omega_{ci}^2}, \tag{10.9}$$

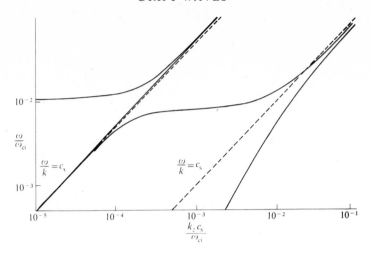

FIG. 10.3. The theoretical dispersion relations for low-frequency waves with the same parameters as Fig. 10.1 but at higher phase velocities or longer wavelengths when coupling with the Alfvén wave also has to be taken into account.

which is shown diagrammatically in Fig. 10.3 and is the zero-ion-temperature limit of a result given by Mikhailovski (1965) ignoring Landau damping

$$\left\{\left(1-\frac{k_y v_0}{\omega}\right)\left(1+i\sqrt{\pi}\,\frac{\omega}{k_z c_e}\right)-\frac{k_z^2 k_B T_e}{M\omega^2}\,I_0(z_{ic})\exp(-z_{ic})\left(1+\frac{k_y T_i v_0}{T_e \omega}\right)\right\}\times$$

$$\times\left\{\omega^2+\frac{k_y T_i \omega v_0}{T_e}-\frac{z_{ic}}{1-I_0(z_{ic})\exp(-z_{ic})}\,k_z^2 c_A^2\right\}=z_{ic}\frac{T_e}{T_i}\,k_z^2 c_A^2\left(1+\frac{k_y T_i v_0}{T_e \omega}\right),$$

where $z_{ic}=k_y^2 k_B T_i/M\omega_{ci}^2$ and finite ion Larmor radius effects have been included.

Eqn (10.9) is a quartic equation in ω for fixed k_y and a given value of k_z, and for large ω and k_z the four branches are asymptotic to the ion sound and Alfvén branches, but for $k_z \to 0$ the branches are

$$\omega \approx \pm k c_A, \qquad \omega \approx -\frac{k_z^2 c_s^2}{k_y v_{0y}}.$$

Experimental work for the phase velocities required by this analysis and under collisionless conditions appear to have been confined to Q-machine plasmas and account has had to be taken of rotation of the plasma column caused by the existence of radial electric fields (see e.g. Little and Middleton 1969).

10.3. The drift dissipative instability

We now turn to a consideration of the situation in which collisional effects predominate in order to show that again waves can propagate and become unstable. Physically in this case the resonance causing instability is between collisional particle motion and the obliquely propagating wave. For simplicity we consider a slab model of the column with the x-direction corresponding to the radial direction. The steady state is supposed to be given by

$$k_B T_e \frac{1}{n_0} \frac{dn_0}{dx} + e \frac{d\phi}{dx} + e v_y B_0 - m v_x \nu_e = 0,$$

$$e v_x B_0 - m v_y \nu_e = 0.$$

Since $v_x = v_y(\nu_e/\omega_{ce})$, for $\nu_e \gg \omega_{ce}$ $v_x \gg v_y$.

If, following the suggestion in § 5.10, we take the radial field to be small $d\phi/dx = 0$ gives

$$-k_B T_e \kappa_n = e B_0 v_y (1 + (\nu_e^2/\omega_{ce}^2)$$

or

$$v_y \approx -\frac{\kappa_n k_B T_e}{e B_0}.$$

The ion motion is given by corresponding equations

$$v_y = -\frac{\kappa_n k_B T_i}{e B_0}, \qquad v_x = -\frac{\nu_i}{\omega_{ci}} v_y,$$

but with $T_i = 0$ both components vanish. The perturbed equations from

$$-k_B T_e \frac{\nabla n_e}{n_e} + e \nabla \phi - e(\mathbf{v}_e \times \mathbf{B}) - m \nu_e v_e = 0$$

are

$$m \nu_e v_{ex1} = i k_x e \phi_1 - k_B T_e (i k_x - \kappa_n) \frac{n_{e1}}{n_0} - e B v_{ey1},$$

$$m \nu_e v_{ey1} = i k_y e \phi_1 - k_B T_e (i k_y) \frac{n_{e1}}{n_0} + e B v_{ex1},$$

$$m \nu_e v_{ez1} = i k_z e \phi_1 - k_B T_e (i k_z) \frac{n_{e1}}{n_0},$$

and the continuity equations gives $\omega n_{e1} = n_0(k_x v_{ex1} + k_y v_{ey1} + k_z v_{ez1})$ with $\nu_e \gg \omega_{ce}$ and $k_x = 0$ we find

$$\frac{n_{e1}}{n_0} = \frac{k_z^2 - i(\nu_e/\omega_{ce})k_y \kappa_n}{(k_z^2 k_B T_e/m\nu_e) - i\omega} \cdot \frac{e}{m\nu_e} \phi_1. \tag{10.10}$$

For the ions

$$M\frac{dv_i}{dt} = -e\nabla\phi + e(\mathbf{v}_i \times \mathbf{B}) - Mv_i\mathbf{v}_i - T_i\frac{\nabla n_i}{n_i}$$

$$-i\omega Mv_{ix1} = -eik_x\phi_1 + ev_{iy1}B - Mv_iv_{ix1} - T_i(ik_x - \kappa_n)$$

$$-i\omega Mv_{iy1} = -eik_y\phi_1 - ev_{ix1}B - Mv_iv_{iy1} - T_i(ik_y)$$

$$-i\omega Mv_{iz1} = -eik_z\phi_1 - Mv_iv_{iz1} - T_i(ik_z)$$

$$i\omega n_{i1} = in_0(k_xv_{ix1} + k_yv_{iy1} + k_zv_{iz1})$$

with

$$T_i = 0, \qquad k_x = 0, \qquad v_i \gg \omega_{ci}$$

gives

$$\frac{n_{i1}}{n_0} = -\frac{e\phi_1}{M\omega_{ci}\omega}\left\{k_y\kappa_n + k_y^2\left(\frac{\omega + iv_i}{\omega_{ci}}\right)\right\}. \qquad (10.11)$$

Quasi-neutrality gives $n_{i1} = n_{e1}$, and the dispersion relation resulting is

$$\omega^2 + i\omega(D_ek_z^2 + v_i + \omega_s) - D_ek_z^2v_i + i\omega_s\omega^* = 0 \qquad (10.12)$$

for $\omega \ll \omega_{ci}$, where $D_e = k_BT_e/mv_e$,

$$\omega^* = k_y\kappa_n\frac{k_BT_e}{eB} \qquad \omega_s = \frac{k_z^2\omega_{ce}\omega_{ci}}{k_y^2v_e}.$$

This has real roots given by the simultaneous solution of

$$\omega^2 = D_ek_z^2v_i \equiv c_s^2\frac{\mu_e}{\mu_i}k_z^2$$

and

$$\omega = \frac{\omega_s\omega^*}{D_ek_z^2 + v_i + \omega_s}.$$

Since $\omega_s \propto k_z^2$, writing

$$\omega = \omega^*\bigg/\left(1 + \frac{v_i}{\omega_s} + \frac{D_ek_z^2}{\omega_s}\right), \qquad (10.13)$$

the values for marginal stability are given by values of ω, as shown in Fig. 10.4 for the curve $B = B_1$. For $\omega_1 < \omega < \omega_2$ the imaginary part of ω is positive and instability results. There will be limiting stability for the curve $B = B_2$, and from eqn (10.13) one has

$$\omega \propto B\bigg/\left(D_ek_z^2 + v_i + \frac{k_z^2e^2B^2}{Mmk_y^2v_e}\right),$$

thus as B increases for fixed k_z, ω at first increases and then decreases, so that there will in general be a range of values of B for which instability occurs with an upper and lower limit.

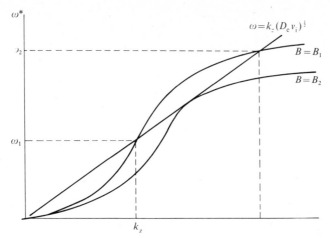

ω^*

$\omega = k_z (D_e v_i)^{\frac{1}{2}}$

ω_2 ---- ---- ---- ---- ---- ---- ---- ---- ---- ---- $B = B_1$

$B = B_2$

ω_1 ---- ----

k_z

FIG. 10.4. Schematic dispersion relation for the drift dissipative instability showing the branch below the drift frequency ω^* and the region of growth $\omega_1 < \omega < \omega_2$ for some magnetic field B_1. For $B < B_2$ all frequencies are damped.

This type of instability was first investigated by Timofeev (1964) and is usually described as the drift dissipative, a comprehensive survey of treatment using kinetic equations has been given by Rukhadze and Silin (1969).

Experimental measurements on an afterglow positive column have been reported by Alcock and Keen (1971), who found it necessary to include the effects of the ion steady-state motion and the wavenumber in the radial direction in order to obtain good numerical agreement. Typical results are shown in Fig. 10.5 for the first-order azimuthal mode in helium at two magnetic fields showing the frequency of naturally occurring oscillations compared with theory for both real and imaginary parts of the frequency. The wavenumber was varied by varying the column length and the other parameters were measured under the experimental conditions.

Measurements of the drift dissipative instability have also been made on a radiofrequency generated plasma by Powers and Mumola (1971), who in their model treat the column as an ambipolar column as in § 3.2, and effectively use the plane solution for an ambipolar column to express the steady-state transverse (radial) variation of number density and suppose that all perturbed quantities vary in the same way in the x-direction thus $n_{e1} \propto \sin(\pi x/x_0)\exp\{i(k_y y + k_z z - \omega t)\}$. The resulting dispersion relation is of the same form as (10.12), but, in addition to those terms included in the treatment of Alcock and Keen, it now has terms arising from the ionization rate and the steady-state electric field. Fig. 10.6 shows a comparison between theory and experiment for helium, the

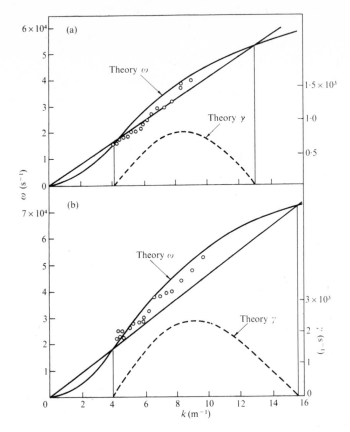

FIG. 10.5. Measured dispersion of drift dissipative waves in a helium plasma column at 20 mTorr, $r_w = 12·5$ mm, at two values of axial magnetic field: (a) 50 mTorr, (b) 35 mTorr. The theoretical region of growth and the corresponding growth rate is shown dashed. The experimental points indicate frequencies for which it was possible to excite waves.

gas in which the instability was first observed by Hoh and Lehnert (1960) in an active discharge. In this case the magnetic field for the onset of instability B_c as the magnetic field is increased under constant plasma conditions is shown plotted in terms of similarity variables $B_c r_w$ versus $p r_w$, r_w being the column radius.

Treatments of drift waves in cylindrical geometry have been given by, for instance, Chen (1967) and Chu, Coppi, Hendel, and Perkins (1969). Additional complications arise because the drift velocity is a function of radius, and thus there is a tendency for there to be localization of the wave amplitude at the radius corresponding to the maximum growth rate for the frequency being excited. This fact renders the plane-geometry

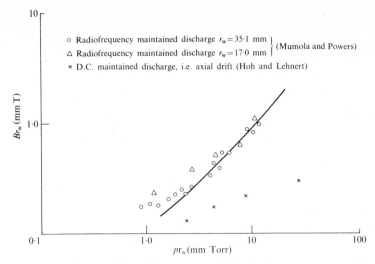

FIG. 10.6. Measurements of the critical magnetic field for the onset of instability in a helium plasma column. The theoretical line is the predicted onset of the drift dissipative instability in the absence of drift.

model applied at the radius corresponding to maximum amplitude a good approximation.

Early theoretical experiments and treatments (Kadomtsev and Nedospasov 1960) were complicated by the combined effects of axial drift due to the current in a d.c. discharge together with the drift dissipative instability, and it is for this reason that experiments turned to radiofrequency and afterglow discharges in order to remove the additional destabilizing effect of synchronism of the drifting particles and a helical perturbation. To bring out the need for this distinction, the results of Hoh and Lehnert in helium have been plotted in Fig. 10.6, showing a generally similar variation but lower threshold field.

The existence of instability above a critical magnetic field affects all the transport coefficients of the background plasma, and it is this which causes the enhanced diffusion (Bohm diffusion, characterized by $D_{am} \propto B^{-1}$) across the magnetic field noted in Fig. 3.6 (p. 45). The determination of appropriate values under such conditions has excited considerable interest (see e.g. Kadomtsev 1965; Simon and Shiau 1969; Monticello and Simon 1974). The process of turbulent diffusion depends on the particular instability which gives rise to the fluctuating fields which are present in the plasma but a general argument can be applied in magnetized plasmas of sufficient length that radial losses are dominant. This rests on supposing that the fluctuating electric fields E_f have a correlation length λ_c which is related to the temperature by $eE_f\lambda_c \sim k_B T_e$. The

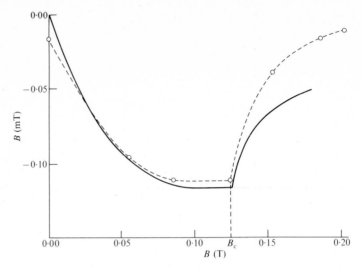

FIG. 10.7. A comparison of theoretical (solid line) and experimental (dashed line) values for the induced magnetic field in a helium plasma column corresponding to $p = 0.6$ mTorr and discharge current 100 mA, $r_w = 15$ mm, showing how the classical diamagnetism is influenced by the onset of the drift dissipative instability at the critical field B_c. (Nagai, Imazu, Maruyama, and Maruyama 1975).

electric field can cause a maximum particle drift v_{Df} given by $v_{Df} = E_f/B$, and this results in a diffusion coefficient D_f which, taking λ_c as the step-length for a random process, gives

$$D_f \lesssim \lambda_c v_{Df} = \frac{k_B T_e}{eB},$$

a result first given by Bohm (1949). For an extensive discussion related to the drift instability see Kadomtsev (1965). A formally similar result is obtained if in the expression for the electron diffusion coefficient D_{em} given in § 3.1 the effective collision frequency ν_e is taken to be the electron cyclotron frequency ω_{ce} and ionization is ignored. This can be justified heuristically if one recognizes that the Debye sphere may be large compared with the electron cyclotron radius.

A recent measurement showing the diamagnetism is also affected by turbulence is given in Fig. 10.7.

APPENDIX I

More general solutions of the Boltzmann equation

THE fundamental quantities in collision processes involving atoms and charged particles are the collision cross-sections. These are usually expressed as a function of energy ε, and writing the cross-section for momentum transfer $q_m(\varepsilon)$ the equation for the distribution function derived from the Boltzmann eqn (1.11) becomes

$$\frac{d}{d\varepsilon}\left(\frac{e^2E^2\varepsilon}{3n_g q_m(\varepsilon)}\frac{df}{d\varepsilon}\right)+\frac{2m}{M}\frac{d}{d\varepsilon}\left\{\varepsilon^2 n_g q_m(\varepsilon)\left(f+k_B T_e\frac{df}{d\varepsilon}\right)\right\}=0.$$

The cases of constant collision frequency and constant mean free path treated in §1.5 correspond to $q_m(\varepsilon)$ constant and $q_m(\varepsilon)\propto\varepsilon^{-\frac{1}{2}}$ respectively.

The inclusion of inelastic processes is straightforward in principle, with additional terms for input and output from inelastic and superelastic collisions (collisions of the second kind), to give

$$\frac{d}{d\varepsilon}\left(\frac{e^2E^2\varepsilon}{3n_g q_m(\varepsilon)}\frac{df}{d\varepsilon}\right)+\frac{2m}{M}\frac{d}{d\varepsilon}\left\{\varepsilon^2 n_g q_m(\varepsilon)\left(f+k_B T_e\frac{df}{d\varepsilon}\right)\right\}+$$

$$+\sum_i\left\{(\varepsilon+\varepsilon_i)f(\varepsilon+\varepsilon_i)n_i q_i(\varepsilon+\varepsilon_i)-\varepsilon f(\varepsilon)n_i q_i(\varepsilon)\right\}+$$

$$+\sum_j\left\{(\varepsilon-\varepsilon_j)f(\varepsilon-\varepsilon_j)n_j q_{-j}(\varepsilon-\varepsilon_j)-\varepsilon f(\varepsilon)n_j q_{-j}(\varepsilon)\right\}=0,$$

$$\text{(AI.1)}$$

where n_i and n_j are the densities of excited levels of excitation energy ε_i, ε_j and the cross-sections have a negative suffix for superelastic processes.

Multiplying by $(2/m)^{\frac{1}{2}}\varepsilon$ and integrating gives

$$-eE^2\left(\frac{2}{m}\right)^{\frac{1}{2}}\frac{e}{3n_g}\int_0^\infty\frac{\varepsilon}{q_m(\varepsilon)}\frac{df}{d\varepsilon}.d\varepsilon$$

$$=\left(\frac{2}{m}\right)^{\frac{1}{2}}\frac{2m}{M}\int_0^\infty\varepsilon^2 n_g q_m(\varepsilon)\left(f+k_B T_e\frac{df}{d\varepsilon}\right)d\varepsilon+$$

$$+\left(\frac{2}{m}\right)^{\frac{1}{2}}\sum_j\varepsilon_j\int_0^\infty\varepsilon f(\varepsilon)\{n_j q_j(\varepsilon)-n_{-j}q_{-j}(\varepsilon)\}\,d\varepsilon.$$

With the mobility μ defined by

$$-\left(\frac{2}{m}\right)^{\frac{1}{2}} \frac{e}{3n_g} \int_0^\infty \frac{\varepsilon}{q_m(\varepsilon)} \frac{df}{d\varepsilon} d\varepsilon \qquad (AI.2)$$

the left-hand side is eEv_D, the energy input.

Given the cross-sections and the densities of excited states, the form of the distribution function for a given energy input can be determined.

From the form of eqn (AI.1) it can readily be seen that the detailed shape of the distribution function will change as each successive excitation potential is exceeded, and this is seen in the work quoted in Chapters 5 and 8.

The diffusion coefficient D is given by

$$\left(\frac{2}{m}\right)^{\frac{1}{2}} \frac{1}{3n_g} \int_0^\infty \frac{\varepsilon f}{q_m(\varepsilon)} d\varepsilon, \qquad (AI.3)$$

with the result that for constant q_m for both Maxwellian and Druyvesteyn distributions the relation

$$D = \frac{\bar{c}\lambda}{3} \qquad (AI.4)$$

holds, where

$$\bar{c} = \frac{\int v g(v) \, dv}{\int g(v) \, dv}$$

is the mean speed, g is the isotropic velocity distribution function, and λ is the mean free path $1/nq_m$.

The mobility μ is given by

$$\mu = \frac{\bar{c}\lambda}{3} \frac{e}{k_B T} \qquad (AI.5)$$

for a Maxwellian distribution and

$$\mu = \sqrt{\pi} \frac{\bar{c}}{3} E \sqrt{\left(\frac{3m}{M}\right)} = \frac{\bar{c}}{E} \left(\frac{\pi m}{3M}\right)^{\frac{1}{2}} \quad \text{or} \quad v_D = \sqrt{\left(\frac{\pi m}{3M}\right)} \bar{c} \qquad (AI.6)$$

for a Druyvesteyn distribution.

If we take the Boltzmann equation and include the effect of collisions between charged particles as well as inelastic processes, but ignore superelastic collisions, (AI.1) will be modified to the form

$$\frac{d}{d\varepsilon_N} \left\{ (\nu_e + \nu_E) \frac{df}{d\varepsilon_N} + (\nu_e + \nu_g) f \right\} = \nu_a f, \qquad (AI.7)$$

where
$$\varepsilon_N = \varepsilon / \bar{\varepsilon}$$

and
$$\nu_e = n_e \frac{e^4}{2\pi\varepsilon_0^2 m^{\frac{1}{2}}\varepsilon^{\frac{3}{2}}} A(\sqrt{\varepsilon_N}) \ln \Lambda,$$

with
$$A(x) = \mathrm{erf}(x) - \frac{2x}{\sqrt{\pi}} \exp(x^2).$$

ln Λ, the Coulomb logarithm, is the electron–electron collision frequency and is directly related to α in eqn (1.13);

$$\nu_E = n_g \frac{4\varepsilon_N}{3_\varepsilon^{-\frac{3}{2}}} \cdot \frac{m^{\frac{3}{2}}}{Q_t} \left(\frac{eE}{mn_g}\right)^2$$

is the field relaxation parameter;

$$\nu_g = \frac{2m}{M} n_g \left(\frac{\varepsilon}{m}\right)^{\frac{1}{2}} \frac{\varepsilon^2}{\bar{\varepsilon}^2} Q_t$$

is the collision frequency for momentum transfer; and

$$\nu_a = n_g Q_a \left(\frac{\bar{\varepsilon}}{m}\right)^{\frac{1}{2}} \left(\frac{\varepsilon}{\bar{\varepsilon}}\right)$$

is the inelastic collision frequency.

Given Q_t and Q_a the elastic and inelastic total collision cross-sections as functions of energy, and taking $\bar{\varepsilon}$, as n_e, n_g, and E parameters, it is possible to compute the electron energy distribution function, and thus estimate more accurately the rates of various atomic or molecular processes. For an atomic gas Q_a will have a threshold ε_{exc} at the first excitation energy, and thus, for sufficiently high densities and low electric fields, the distribution function will be Maxwellian in form for ε below ε_{exc}. This corresponds to the ν_E, ν_g, $\nu_a \to 0$ limit of eqn (AI.7). On the other hand, well above ε_{exc} at densities such that $\nu_a \gg \nu_e$ we will have a departure from the Maxwellian form. One can conclude that at low densities the form of the distribution will be approximately Druyvesteyn, that at high densities ($\nu_e > \nu_a$) it will be approximately Maxwellian, and that in the intermediate range of densities which corresponds to the majority of discharges it will lie between the two. The effect of inelastic processes can most readily be exemplified by considering the special case ν_E, $\nu_g \ll \nu_e$, when the solution for the distribution function is readily found to be

$$f(\varepsilon) \propto \exp(-\varepsilon / k_B T_e) \qquad \varepsilon < \varepsilon_{exc}$$

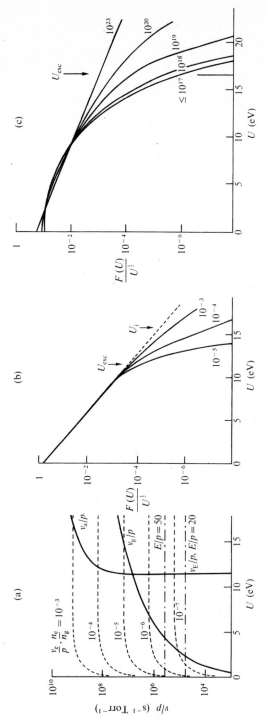

FIG. AI.1. Computations of distribution functions in rare gases. (a) shows the relative importance of the different collision terms in eqn (AI.7) for argon. The electron–electron collision frequency ν_e is given for different degrees of ionization n_e/n_g; the field relaxation term ν_E is shown for two reduced fields E/p in V m^{-1} Torr^{-1}; the inelastic collision term is zero below the threshold energy for inelastic processes $\varepsilon_{exc} \equiv 11 \cdot 5$ eV. The electron mean energy $\bar{\varepsilon}$ is taken to be $1 \cdot 44$ eV a typical value for $pr_w = 100$ mm Torr. (b) The computed distribution function for argon with the fractional ionization as a parameter and $\bar{\varepsilon} = 1 \cdot 44$ eV. (c) The results of a similar calculation for neon with mean energy 3 eV, $E = 4 \cdot 65 \times 10^3$ V m^{-1}, $n_g = 3 \times 10^{24}$ m^{-3} (i.e. $\sim$90 Torr), and electron densities ranging from 10^{23} m^{-3} to 10^{17} m^{-3} or less, showing the transition from Maxwellian to Druyvesteyn forms.

and

$$f(\varepsilon) \propto \frac{\varepsilon_{\text{exc}}}{k_{\text{B}} T_e} \frac{\nu_{\text{a}}}{\nu_{\text{e}}} \cdot \exp\left\{ -\frac{\varepsilon}{k_{\text{B}} T_e} \left(1 + \frac{\nu_{\text{a}}}{\nu_e} \right) \right\} \qquad \varepsilon > \varepsilon_{\text{exc}}$$

for $\nu_{\text{a}} < \nu_{\text{e}}$. Thus the high-energy tail above the excitation potential is deficient in electrons relative to a Maxwellian with a temperature given by the main body and can be regarded as being characterized by a different and lower temperature.

This concept has been carried further by Vriens (1973), who proposed explicitly a 'two-temperature' model, where the 'temperature' of the high-energy tail is electron-number-density dependent. Use of this model is referred to in connection with the contraction of the positive column at the end of §5.8. More generally, recourse has to be had to numerical methods unless particular analytical forms are assumed for the cross-sections; a suitable choice leads to the form for the excitation and ionization rates used in eqn (5.9).

Figs AI.1(a), (b), and (c) give computed results for two different cases which bring out the points made above. In Fig. AI.1(b), due to Wojaczek (1967), conditions are such that for $\varepsilon < \varepsilon_{\text{exc}}$ the distribution is Maxwellian, while in Fig. AI.1(c) at low densities the Druvesteyn form is found (van Duyn, Prins, and Smit, 1975).

APPENDIX II

The permittivity tensor for a cold magnetized plasma

CONSIDER electrons only to be mobile for the moment, their equation of motion is

$$\frac{m\,d\mathbf{v}}{dt} = -e(\mathbf{E} + \mathbf{v} \times \mathbf{B}).$$

When the z-axis is taken to coincide with the steady magnetic field so that $\mathbf{B} = B_0\mathbf{k}$, the equation of motion becomes in matrix form

$$-mi\omega \begin{bmatrix} v_x \\ v_y \\ v_z \end{bmatrix} = -e \begin{bmatrix} E_x + v_y B_0 \\ E_y - v_x B_0 \\ E_z \end{bmatrix},$$

if all quantities vary as $\exp(-i\omega t)$. Solving for $\mathbf{E}$ gives

$$\mathbf{E} = \begin{bmatrix} +\dfrac{im\omega}{e} & -\dfrac{m\omega_{ce}}{e} & 0 \\[2mm] \dfrac{m\omega_{ce}}{e} & +\dfrac{im\omega}{e} & 0 \\[2mm] 0 & 0 & +\dfrac{im\omega}{e} \end{bmatrix} \mathbf{v} \equiv \frac{im\omega}{e}\mathbf{K}\mathbf{v}.$$

This inverts to give

$$\mathbf{v} = \frac{ie}{m} \begin{bmatrix} \dfrac{-\omega}{\omega^2 - \omega_{ce}^2} & \dfrac{i\omega_{ce}}{\omega^2 - \omega_{ce}^2} & 0 \\[2mm] \dfrac{-i\omega_{ce}}{\omega^2 - \omega_{ce}^2} & \dfrac{-\omega}{\omega^2 - \omega_{ce}^2} & 0 \\[2mm] 0 & 0 & -\dfrac{1}{\omega} \end{bmatrix} \mathbf{E}.$$

Substituting in Maxwell's equation,

$$\nabla \times \mathbf{H} = \varepsilon_0 \frac{\partial \mathbf{E}}{\partial t} + \mathbf{J},$$

where

$$\mathbf{J} = -ne\mathbf{v},$$

gives

$$\nabla \times \mathbf{H} = \varepsilon_0 \boldsymbol{\varepsilon} \frac{\partial \mathbf{E}}{\partial t},$$

where

$$\boldsymbol{\varepsilon} = \begin{bmatrix} 1 - \dfrac{\omega_{pe}^2}{\omega^2 - \omega_{ce}^2} & +\dfrac{i\omega_{pe}^2 \omega_{ce}}{\omega(\omega^2 - \omega_{ce}^2)} & 0 \\[3ex] -\dfrac{i\omega_{pe}^2 \omega_{ce}}{\omega(\omega^2 - \omega_{ce}^2)} & 1 - \dfrac{\omega_{pe}^2}{\omega^2 - \omega_{ce}^2} & 0 \\[3ex] 0 & 0 & 1 - \dfrac{\omega_{pe}^2}{\omega^2} \end{bmatrix}. \qquad \text{(AII.1)}$$

The right-hand side can be written formally

$$\begin{bmatrix} \varepsilon_{11} & +i\varepsilon_{12} & 0 \\ -i\varepsilon_{12} & \varepsilon_{11} & 0 \\ 0 & 0 & \varepsilon_{33} \end{bmatrix}$$

The matrix $\boldsymbol{\varepsilon}$ is Hermitian and has real eigenvalues which are readily found by solving $\det(\boldsymbol{\varepsilon} - \lambda\mathbf{I}) = 0$ to be $\varepsilon_{11} \mp \varepsilon_{12}$, ε_{33}. The corresponding eigenvectors are $E_x \pm iE_y$, E_z and thus for circularly polarized electric vectors there are effective relative permittivities

$$\varepsilon_l = 1 - \frac{\omega_{pe}^2}{\omega(\omega + \omega_{ce})}, \qquad \varepsilon_r = 1 - \frac{\omega_{pe}^2}{\omega(\omega - \omega_{ce})}. \qquad \text{(AII.2)}$$

The addition of ions generalizes these to

$$\varepsilon_l = 1 - \frac{\omega_{pe}^2}{\omega(\omega + \omega_{ce})} - \frac{\omega_{pi}^2}{\omega(\omega - \omega_{ci})}, \qquad \varepsilon_r = 1 - \frac{\omega_{pe}^2}{\omega(\omega - \omega_{ce})} - \frac{\omega_{pi}^2}{\omega(\omega + \omega_{ci})}, \qquad \text{(AII.3)}$$

while

$$\varepsilon_{11} = 1 - \frac{\omega_{pe}^2}{\omega^2 - \omega_{ce}^2} - \frac{\omega_{pi}^2}{\omega^2 - \omega_{ci}^2}$$

and

$$\varepsilon_{12} = \frac{\omega_{pe}^2 \omega_{ce}}{\omega(\omega^2 - \omega_{ce}^2)} - \frac{\omega_{pi}^2 \omega_{ci}}{\omega(\omega^2 - \omega_{ci}^2)}. \qquad \text{(AII.4)}$$

The inclusion of collisional terms in the equations of motion causes modification to the extent that the permittivity matrix is no longer

Hermitian and the terms become

$$\varepsilon_{11} = 1 - \frac{\omega_{pe}^2(\omega + i\nu_e)}{\omega\{(\omega + i\nu_e)^2 - \omega_{ce}^2\}} - \frac{\omega_{pi}^2(\omega + i\nu_i)}{\omega\{(\omega + i\nu_i)^2 - \omega_{ci}^2\}},$$

$$\varepsilon_{12} = + \frac{\omega_{pe}^2\omega_{ce}}{\omega\{(\omega + i\nu_e)^2 - \omega_{ce}^2\}} - \frac{\omega_{pi}^2\omega_{ci}}{\omega\{(\omega + i\nu_i)^2 - \omega_{ci}^2\}},$$

$$\varepsilon_{33} = 1 - \frac{\omega_{pe}^2}{\omega(\omega + i\nu_e)} - \frac{\omega_{pi}^2}{\omega(\omega + i\nu_i)}.$$

Or separating real and imaginary parts and introducing corresponding subscripts r and i

$$\varepsilon_{11r} = 1 - \frac{\omega_{pe}^2(\omega^2 + \nu_e^2 - \omega_{ce}^2)}{(\omega^2 - \nu_e^2 - \omega_{ce}^2)^2 + 4\omega^2\nu_e^2} - \frac{\omega_{pi}^2(\omega^2 + \nu_i^2 - \omega_{ci}^2)}{(\omega^2 - \nu_i^2 - \omega_{ci}^2)^2 + 4\omega^2\nu_i^2},$$

$$\varepsilon_{11i} = + \frac{\omega_{pe}^2\nu_e(\omega^2 + \nu_e^2 + \omega_{ce}^2)}{\omega\{(\omega^2 - \nu_e^2 - \omega_{ce}^2)^2 + 4\omega^2\nu_e^2\}} + \frac{\omega_{pi}^2\nu_i(\omega^2 + \nu_i^2 + \omega_{ci}^2)}{\omega\{(\omega^2 - \nu_i^2 - \omega_{ci}^2) + 4\omega^2\nu_i^2\}},$$

$$\varepsilon_{12r} = + \frac{\omega_{ce}\omega_{pi}^2(\omega^2 - \nu^2 - \omega_{ce}^2)}{\omega\{(\omega^2 - \nu_e^2 - \omega_{ce}^2)^2 + 4\omega^2\nu_e^2\}} - \frac{\omega_{ci}\omega_{pi}^2(\omega^2 - \nu_i^2 - \omega_{ci}^2)}{\omega\{(\omega^2 - \nu_i^2 - \omega_{ci}^2)^2 + 4\omega^2\nu_i^2\}},$$

$$\varepsilon_{12i} = 2\left\{ -\frac{\omega_{pe}^2\omega_{ce}\nu_e}{(\omega^2 - \nu_e^2 - \omega_{ce}^2)^2 + 4\omega^2\nu_e^2} + \frac{\omega_{pi}^2\omega_{ci}\nu_i}{(\omega^2 - \nu_i^2 - \omega_{ci}^2)^2 + 4\omega^2\nu_i^2} \right\},$$

$$\varepsilon_{33r} = 1 - \frac{\omega_{pe}^2}{\omega^2 + \nu_e^2} - \frac{\omega_{pi}^2}{\omega^2 + \nu_i^2},$$

$$\varepsilon_{33i} = + \frac{\omega_{pe}^2\nu_e}{\omega(\omega^2 + \nu_e^2)} + \frac{\omega_{pi}^2\nu_i}{\omega(\omega^2 + \nu_i^2)}.$$

For some purposes it is convenient to recast the problem again in terms of the in-phase and out-of-phase currents, i.e. conduction and displacement currents. Thus

$$\varepsilon_0\varepsilon\frac{\partial \mathbf{E}}{\partial t} \equiv \varepsilon_0(\varepsilon_r + i\varepsilon_i)\frac{\partial \mathbf{E}}{\partial t} \equiv -i\omega\varepsilon_0(\varepsilon_r + i\varepsilon_i)\mathbf{E},$$

and the last term is written $\mathbf{J} = \boldsymbol{\sigma}\mathbf{E}$ or inversely $\mathbf{E} = \boldsymbol{\eta}\mathbf{J}$, so that

$$\boldsymbol{\sigma} \equiv +\omega\varepsilon_0\varepsilon_i = \boldsymbol{\eta}^{-1}, \tag{AII.5}$$

defining the conductivity and resistivity tensors.

One can examine the form of the equations in various limiting cases

depending on the relative values of the frequencies involved. Four such are:

(1) $\omega < \nu_i,\ \nu_e < \omega_{ci},\ \omega_{ce}$ low-frequency, collisional, strongly magnetized;

(2) $\omega_{ci},\ \omega_{ce} < \nu_i,\ \nu_e < \omega$ high-frequency, collisionless, unmagnetized;

(3) $\nu_i,\ \nu_e < \omega < \omega_{ci},\ \omega_{ce}$ strongly magnetized, collisionless, intermediate frequency;

(4) $\nu_i,\ \nu_e < \omega_{ci} < \omega < \omega_{ce}$ collisionless, intermediate frequency;

with corresponding forms

(1) $$\varepsilon_{11} = 1 + \frac{i\nu_e\omega_{pe}^2}{\omega\omega_{ce}^2} + \frac{i\nu_i\omega_{pi}^2}{\omega\omega_{ci}^2},$$

$$\varepsilon_{12} = -\frac{\omega_{pi}^2\omega_{ci}\nu_e^2 + \omega_{pe}^2\omega_{ce}\nu_i^2}{\omega\omega_{ce}^2\omega_{ci}^2}, \qquad \varepsilon_{33} = 1 - \frac{\omega_{pe}^2}{\omega\nu_e} - \frac{\omega_{pi}^2}{\omega\nu_i};$$

(2) $$\varepsilon_{11} = \varepsilon_{33} = 1 - \frac{\omega_{pe}^2}{\omega^2} - \frac{\omega_{pi}^2}{\omega^2} + \frac{i\omega_{pe}^2\nu_e}{\omega^3} + \frac{i\omega_{pi}^2\nu_i}{\omega^3}, \qquad \varepsilon_{12} = 0;$$

(3) $$\varepsilon_{11} = 1 + \frac{\omega_{pe}^2}{\omega_{ce}^2} + \frac{\omega_{pi}^2}{\omega_{ci}^2}, \qquad \varepsilon_{12} = -\frac{\omega\omega_{pe}^2}{\omega_{ce}\omega_{ci}^2}, \qquad \varepsilon_{33} = 1 - \frac{\omega_{pe}^2}{\omega^2} - \frac{\omega_{pi}^2}{\omega^2};$$

(4) $$\varepsilon_{11} = 1 + \frac{\omega_{pe}^2}{\omega_{ce}^2} - \frac{\omega_{pi}^2}{\omega^2}, \qquad \varepsilon_{12} = +\frac{\omega_{pe}^2}{\omega\omega_{ce}}, \qquad \omega_{33} = 1 - \frac{\omega_{pe}^2}{\omega^2} - \frac{\omega_{pi}^2}{\omega^2};$$

The transformation into cylindrical coordinates can readily be carried out. Taking the equation for the electric displacement $\mathbf{D} = \varepsilon\mathbf{E}$ in (x, y, z) coordinates, where

$$\varepsilon = \begin{bmatrix} \varepsilon_{11} & +i\varepsilon_{12} & 0 \\ -i\varepsilon_{12} & \varepsilon_{11} & 0 \\ 0 & 0 & \varepsilon_{33} \end{bmatrix}$$

and introducing cylindrical polar coordinates (r, θ, z), one can write $\mathbf{D}' = \mathbf{R}\mathbf{D}$, $\mathbf{E}' = \mathbf{R}\mathbf{E}$, with

$$\mathbf{R} = \begin{bmatrix} \cos\theta & \sin\theta & 0 \\ -\sin\theta & \cos\theta & 0 \\ 0 & 0 & 1 \end{bmatrix}$$

so that $\mathbf{D}' = \mathbf{R}\boldsymbol{\varepsilon}\mathbf{R}^{-1}\mathbf{E}'$, whence the transformed permittivity matrix is found to be

$$
\begin{bmatrix}
\varepsilon_{11} & +i\varepsilon_{12} & 0 \\
-i\varepsilon_{12} & \varepsilon_{11} & 0 \\
0 & 0 & \varepsilon_{33}
\end{bmatrix}
$$

i.e. $\boldsymbol{\varepsilon}' = \boldsymbol{\varepsilon}$.

In the electrostatic approximation $\mathbf{E} = -\nabla\phi$, so that with azimuthal symmetry

$$
\mathbf{E}' =
\begin{bmatrix}
-\dfrac{\partial\phi}{\partial r} \\[2mm]
-\dfrac{\partial\phi}{\partial z}
\end{bmatrix}
$$

giving

$$
\mathbf{D}' = -
\begin{bmatrix}
\varepsilon_{11} & \dfrac{\partial\phi}{\partial r} \\[2mm]
-i\varepsilon_{12} & \dfrac{\partial\phi}{\partial r} \\[2mm]
\varepsilon_{33} & \dfrac{\partial\phi}{\partial z}
\end{bmatrix}
$$

and the wave equation follows from $\nabla \cdot \mathbf{D}' = 0$, i.e.

$$
\frac{1}{r}\frac{\partial}{\partial r}\left(\varepsilon_{11} r \frac{\partial\phi}{\partial r}\right) + \varepsilon_{33}\frac{\partial^2\phi}{\partial z^2} = 0. \tag{AII.6}
$$

This form then is applicable to azimuthally symmetric electrostatic waves. The more general case allowing phase velocities comparable with c, the velocity of light in free space, is given in Appendix III for cold plasma.

APPENDIX III

Wave propagation in cylindrical magnetized plasmas

WE HAVE seen that in the absence of collisions the natural coordinates for wave propagation are the magnetic field direction z and clockwise and counterclockwise rotating vectors in the x, y plane. There is considerable simplification and elegance of presentation to be gained when dealing with problems with cylindrical symmetry and an axial magnetic field to use such vectors as a basis.

Defining l and r by

$$l = \frac{x - iy}{\sqrt{2}} \quad \text{and} \quad r = \frac{x + iy}{\sqrt{2}}$$

the coordinate system $\mathbf{l} \equiv (l, r, z)$ and $\mathbf{x} \equiv (x, y, z)$ are related by $\mathbf{l} = U\mathbf{x}$, where

$$U = \frac{1}{\sqrt{2}} \begin{bmatrix} 1 & -i & 0 \\ 1 & i & 0 \\ 0 & 0 & \sqrt{2} \end{bmatrix} \tag{AIII.1}$$

and the vector operations of multiplications and differentiation can be shown to be given by

$$\mathbf{A} \cdot \mathbf{B} \equiv A_x B_x + A_y B_y + A_z B_z = A_l B_r + A_r B_l + A_z B_z \tag{AIII.2}$$

$$\mathbf{A} \times \mathbf{B} \equiv \begin{vmatrix} \mathbf{i} & \mathbf{j} & \mathbf{k} \\ A_x & A_y & A_z \\ B_x & B_y & B_z \end{vmatrix} = i \begin{vmatrix} \hat{\mathbf{l}} & \hat{\mathbf{r}} & \mathbf{k} \\ A_r & A_l & A_z \\ B_r & B_l & B_z \end{vmatrix} \tag{AIII.3}$$

$\hat{}$ denoting unit vectors, such that

$$\hat{\mathbf{l}} = \frac{1}{\sqrt{2}} (\mathbf{i} + i\mathbf{j}), \qquad \hat{\mathbf{r}} = \frac{1}{\sqrt{2}} (\mathbf{i} - i\mathbf{j}),$$

or

$$\mathbf{i} = \frac{1}{\sqrt{2}} (\hat{\mathbf{l}} + \hat{\mathbf{r}}), \qquad \mathbf{j} = \frac{i}{\sqrt{2}} (\hat{\mathbf{r}} - \hat{\mathbf{l}}).$$

$$\nabla \phi \equiv \mathbf{i} \frac{\partial \phi}{\partial x} + \mathbf{j} \frac{\partial \phi}{\partial y} + \mathbf{k} \frac{\partial \phi}{\partial z}$$

transforms to

$$\hat{\mathbf{l}}\frac{\partial\phi}{\partial r}+\hat{\mathbf{r}}\frac{\partial\phi}{\partial l}+\mathbf{k}\frac{\partial\phi}{\partial z}\,.\qquad\text{(AIII.4)}$$

And caution is needed in operating with

$$\nabla_{lrz}\equiv\hat{\mathbf{l}}\frac{\partial}{\partial r}+\hat{\mathbf{r}}\frac{\partial}{\partial l}+\mathbf{k}\frac{\partial}{\partial z}\,,$$

since, while $\hat{\mathbf{l}}.\mathbf{k}=\hat{\mathbf{r}}.\mathbf{k}=0$, $\hat{\mathbf{l}}.\hat{\mathbf{r}}=1$ and $\hat{\mathbf{l}}.\hat{\mathbf{l}}=0=\hat{\mathbf{r}}.\hat{\mathbf{r}}$, so that

$$\nabla.\mathbf{A}\equiv\frac{\partial A_r}{\partial r}+\frac{\partial A_l}{\partial l}+\frac{\partial A_z}{\partial z}\qquad\text{(AIII.5)}$$

and furthermore,

$$\hat{\mathbf{l}}\times\hat{\mathbf{r}}=-\mathrm{i}\mathbf{k},\qquad\hat{\mathbf{r}}\times\mathbf{k}=-\mathrm{i}\hat{\mathbf{r}},\qquad\hat{\mathbf{k}}\times\hat{\mathbf{l}}=-\mathrm{i}\hat{\mathbf{l}},$$

so that

$$\nabla\times\mathbf{A}\equiv\left(\hat{\mathbf{l}}\frac{\partial}{\partial r}+\hat{\mathbf{r}}\frac{\partial}{\partial l}+\mathbf{k}\frac{\partial}{\partial z}\right)\times(A_l\hat{\mathbf{l}}+A_r\hat{\mathbf{r}}+A_z\hat{\mathbf{k}})$$

$$=\mathrm{i}\left\{\hat{\mathbf{r}}\left(\frac{\partial A_r}{\partial z}-\frac{\partial A_z}{\partial l}\right)+\hat{\mathbf{l}}\left(\frac{\partial A_z}{\partial r}-\frac{\partial A_l}{\partial z}\right)+\mathbf{k}\left(\frac{\partial A_l}{\partial l}-\frac{\partial A_r}{\partial r}\right)\right\}$$

$$\equiv\mathrm{i}\begin{vmatrix}\hat{\mathbf{l}}&\hat{\mathbf{r}}&\mathbf{k}\\[6pt]\dfrac{\partial}{\partial l}&\dfrac{\partial}{\partial r}&\dfrac{\partial}{\partial z}\\[6pt]A_r&A_l&A_z\end{vmatrix}$$

Now we can write out the Maxwell's equations

$$\nabla.(\boldsymbol{\varepsilon}\mathbf{E})=0,\qquad\nabla.\mathbf{H}=0,\qquad\nabla\times\mathbf{E}=\mathrm{i}\omega\mu_0\mathbf{H},\qquad\nabla\times\mathbf{H}=-\mathrm{i}\omega\varepsilon_0\boldsymbol{\varepsilon}\mathbf{E},$$

in (r, l, z) coordinates with components of E and H varying as $A(r, l)\exp(\mathrm{i}kk_0z)$, where k is the wavenumber normalized to the free-space wavenumber of the same frequency,

$$\varepsilon_l\frac{\partial E_l}{\partial l}+\varepsilon_r\frac{\partial E_r}{\partial r}+\mathrm{i}\varepsilon_{33}kk_0E_z=0,\qquad\text{(AIII.7)}$$

$$\frac{\partial H_l}{\partial l}+\frac{\partial H_r}{\partial r}+\mathrm{i}kk_0H_z=0,\qquad\text{(AIII.8)}$$

$$-\mathrm{i}kk_0E_l+\frac{\partial E_z}{\partial r}=+\omega\mu_0H_l\qquad\text{(AIII.9)}$$

$$\frac{\partial E_z}{\partial l}+\mathrm{i}kk_0E_r=+\omega\mu_0H_r,\qquad\text{(AIII.10)}$$

$$\frac{\partial E_l}{\partial l} - \frac{\partial E_r}{\partial r} = +\omega\mu_0 H_z, \tag{AIII.11}$$

$$-ikk_0 H_l + \frac{\partial H_z}{\partial r} = -\omega\varepsilon_0\varepsilon_l E_l, \tag{AIII.12}$$

$$-\frac{\partial H_z}{\partial l} + ikk_0 H_r = -\omega\varepsilon_0\varepsilon_r E_r, \tag{AIII.13}$$

$$\frac{\partial H_l}{\partial l} - \frac{\partial H_r}{\partial r} = -\omega\varepsilon_0\varepsilon_{33}E_z. \tag{AIII.14}$$

Eqns (AIII.7) and (AIII.11), together with (AIII.8) and (AIII.14), give

$$\frac{\partial}{\partial r}\begin{bmatrix} E_r \\ H_r \end{bmatrix} = \begin{bmatrix} -\dfrac{i\varepsilon_{33}kk_0}{\varepsilon_l+\varepsilon_r} & -\dfrac{\varepsilon_l\omega\mu_0}{\varepsilon_l+\varepsilon_r} \\ +\dfrac{\omega\varepsilon_0\varepsilon_{33}}{2} & -\dfrac{ikk_0}{2} \end{bmatrix}\begin{bmatrix} E_z \\ H_z \end{bmatrix} \tag{AIII.15}$$

$$\frac{\partial}{\partial l}\begin{bmatrix} E_l \\ H_l \end{bmatrix} = \begin{bmatrix} -\dfrac{i\varepsilon_{33}kk_0}{\varepsilon_l+\varepsilon_r} & +\dfrac{\varepsilon_r\omega\mu_0}{\varepsilon_l+\varepsilon_r} \\ -\dfrac{\omega\varepsilon_0\varepsilon_{33}}{2} & -\dfrac{ikk_0}{2} \end{bmatrix}\begin{bmatrix} E_z \\ H_z \end{bmatrix} \tag{AIII.16}$$

Eqns (AIII.9) and (AIII.12) give

$$\frac{\partial}{\partial r}\begin{bmatrix} E_z \\ H_z \end{bmatrix} = \begin{bmatrix} +ikk_0 & +\omega\mu_0 \\ -\omega\varepsilon_0\varepsilon_l & +ikk_0 \end{bmatrix}\begin{bmatrix} E_l \\ H_l \end{bmatrix} \tag{AIII.17}$$

and (AIII.10) and (AIII.13) give

$$\frac{\partial}{\partial l}\begin{bmatrix} E_z \\ H_z \end{bmatrix} = \begin{bmatrix} +ikk_0 & -\omega\mu_0 \\ +\omega\varepsilon_0\varepsilon_r & +ikk_0 \end{bmatrix}\begin{bmatrix} E_r \\ H_r \end{bmatrix} \tag{AIII.18}$$

Finally, (AIII.15) and (AIII.17) and (AIII.16) and (AIII.18) give the same equation for

$$\begin{bmatrix} E_l \\ H_l \end{bmatrix}\begin{bmatrix} E_r \\ H_r \end{bmatrix}\begin{bmatrix} E_z \\ H_z \end{bmatrix}$$

namely

$$\left\{2\frac{\partial^2}{\partial l\,\partial r}\mathbf{I} + \begin{bmatrix} a & -b \\ -d & c \end{bmatrix}\right\}\begin{bmatrix} E \\ H \end{bmatrix} = 0,$$

with

$$a = \frac{k_0^2 \varepsilon_{33}(\varepsilon_l + \varepsilon_r - 2k^2)}{\varepsilon_l + \varepsilon_r}, \qquad b = \frac{i\omega\mu_0 k k_0 (\varepsilon_r - \varepsilon_l)}{\varepsilon_l + \varepsilon_r},$$

$$c = \frac{k_0^2 \{k^2(\varepsilon_l + \varepsilon_r) - 2\varepsilon_l \varepsilon_r\}}{\varepsilon_l + \varepsilon_r}, \qquad d = \frac{i\omega\varepsilon_0 k_0 k(\varepsilon_r - \varepsilon_l)\varepsilon_{33}}{\varepsilon_l + \varepsilon_r}.$$

Thus E_z is given by the linear combination $a_1 E_{z1} + a_2 E_{z2}$, where

$$2\frac{\partial^2}{\partial l\,\partial r} E_{zi} + p_i^2 E_{zi} \equiv \frac{\partial^2}{\partial x^2} E_{zi} + \frac{\partial^2}{\partial y^2} E_{zi} + p_i^2 E_{zi} = 0, \quad \text{(AIII.19)}$$

and p_i satisfies

$$p_i^4 - p_i^2(a + c) + ac - bd = 0. \quad \text{(AIII.20)}$$

Hence

$$E_z = \{A_1 J_m(p_1\rho) + A_2 J_m(p_2\rho) + B_1 Y_m(p_1\rho) + B_2 Y_m(p_2\rho)\}e^{-im\theta},$$

where A_1, A_2, B_1, B_2 are constant and the radial coordinate is written ρ to avoid confusion. Note that the azimuthally dependent case is included since $e^{-im\theta} \equiv (l/r)^{m/2}$. The magnetic field H_z is given by

$$H_z = \{A_1 h_1 J_m(p_1\rho) + A_2 h_2 J_m(p_2\rho) + B_1 h_1 Y_m(p_1\rho) + B_2 h_2 Y_m(p_2\rho)\}e^{-im\theta}$$

with

$$h_1 = \frac{a - p_1^2}{b}, \qquad h_2 = \frac{a - p_2^2}{b}.$$

The transverse field components are given by

$$
\begin{bmatrix} E_\rho \\[1em] E_\theta \\[1em] H_\rho \\[1em] H_\theta \end{bmatrix}
= \frac{1}{k_0^2(\varepsilon_l - k^2)(\varepsilon_r - k^2)}
\begin{bmatrix}
+ik_0 k\varepsilon_a & -k_0 k\varepsilon_b & +\omega\mu_0\varepsilon_b & +i\omega\mu_0\varepsilon_a \\[0.8em]
+k_0 k\varepsilon_b & +ik_0 k\varepsilon_a & -i\omega\mu_0\varepsilon_a & +\omega\mu_0\varepsilon_b \\[0.8em]
+\omega\varepsilon_0 k^2\varepsilon_b & -i\omega\varepsilon_0\varepsilon_c & +ik_0 k\varepsilon_a & -k_0 k\varepsilon_b \\[0.8em]
+i\omega\varepsilon_0\varepsilon_c & +\omega\varepsilon_0 k^2\varepsilon_b & +k_0 k\varepsilon_b & +ik_0 k\varepsilon_a
\end{bmatrix}
\begin{bmatrix} \dfrac{\partial E_z}{\partial\rho} \\[1em] \dfrac{1}{\rho}\dfrac{\partial E_z}{\partial\theta} \\[1em] \dfrac{\partial H_z}{\partial\rho} \\[1em] \dfrac{1}{\rho}\dfrac{\partial H_z}{\partial\theta} \end{bmatrix}
$$

$$\text{(AIII.21)}$$

where $\varepsilon_a = \varepsilon_l + \varepsilon_r - 2k^2$, $\varepsilon_b = \varepsilon_l - \varepsilon_r$, $\varepsilon_c = 2\varepsilon_l\varepsilon_r - k^2(\varepsilon_l + \varepsilon_r)$.

For an infinitely conducting cylindrical waveguide E_z and E_θ must vanish at the outer boundary $\rho = \rho_w$ and the coefficients B_1 and B_2 are zero for finite fields at $\rho = 0$. Thus

$$A_1 J_m(p_1 \rho_w) + A_2 J_m(p_2 \rho_w) = 0,$$

$$A_1\{k_0 k \varepsilon_b p_1 J'_m(p_1 \rho_w) + m k_0 k \varepsilon_0 J_m(p_1 \rho_w) -$$
$$- i\omega\mu_0 \varepsilon_a h_1 p_1 J'_m(p_1 \rho_w) - im\omega\mu_0 \varepsilon_b h_1 J_m(p_1 \rho_w)\} +$$
$$+ A_2[k_0 k \varepsilon_b p_2 J'_m(p_2 \rho_w) + m k_0 k \varepsilon_0 J_m(p_2 \rho_w) -$$
$$- i\omega\mu_0 \varepsilon_a h_2 p_2 J'_m(p_2 \rho_w) - im\omega\mu_0 \epsilon_b h_2 J_m(p_2 \rho_w)\} = 0.$$

The determinant gives the dispersion relation which for $m = 0$ is

$$k_0 k \varepsilon_0 \{p_1 J'_m(p_1 \rho_w) J_m(p_2 \rho_w) - p_2 J'_m(p_2 \rho_w) J_m(p_1 \rho_w)\}$$
$$= i\omega\mu_0 \varepsilon_a \{h_1 p_1 J'_m(p_1 \rho_w) J_m(p_2 \rho_w) - h_2 p_2 J'_m(p_2 \rho_w) J_m(p_1 \rho_w)\},$$

and this expression contains the solutions for all axially propagating waves in a uniform plasma in a waveguide, and thus contains in principle many of the results obtained in Chapters 6, 7, and 9 as special cases. Application of the methods of this section to coaxial plasmas has been made by Franklin and Oldfield (1969).

Wave propagation in warm plasmas

IF THE effects of particle pressure, i.e. particle temperature, are included in the equations of motion, considerable complication results.

The momentum equations become

$$-i\omega m \mathbf{v}_e = -e(\mathbf{E} + \mathbf{v}_e \times \mathbf{B}) - \frac{1}{n_0} \nabla(n_{e1} k_B T_e)$$

and

$$-i\omega M \mathbf{v}_i = e(\mathbf{E} + \mathbf{v}_i \times \mathbf{B}) - \frac{1}{n_0} \nabla(n_{i1} k_B T_e).$$

To relate the perturbed number densities and velocities, we need the continuity equations

$$-i\omega n_{e1} + \nabla \cdot (n_0 \mathbf{v}_e) = 0, \qquad -i\omega n_{i1} + \nabla \cdot (n_0 \mathbf{v}_i) = 0$$

or

$$\frac{n_{e1}}{n_0} = \frac{\mathbf{k} \cdot \mathbf{v}_e}{\omega}, \qquad \frac{n_{i1}}{n_0} = \frac{\mathbf{k} \cdot \mathbf{v}_i}{\omega},$$

all quantities varying as $\exp\{i(-\omega t + \mathbf{k} \cdot \mathbf{r})\}$; this gives equations of the form

$$-i\omega m v_{ex} = -e(E_x + v_{ey} B_0) - i k_x (\mathbf{k} \cdot \mathbf{v}_e) \frac{k_B T_e}{\omega}$$

or

$$\mathbf{E} = +\frac{i\omega m}{e} \left\{ \mathbf{K}_e \mathbf{v}_e - (\mathbf{k}\mathbf{k}) \cdot \mathbf{v}_e \frac{c_e^2}{\omega^2} \right\} \qquad (AIV.1)$$

$$= \frac{-i\omega M}{e} \left\{ \mathbf{K}_i \mathbf{v}_i - (\mathbf{k}\mathbf{k}) \cdot \mathbf{v}_i \frac{c_i^2}{\omega^2} \right\}. \qquad (AIV.2)$$

In principle these can be solved for $\mathbf{v}_e$ and $\mathbf{v}_i$ in terms of $\mathbf{E}$, and then Maxwell's equations, in the form

$$k^2 \mathbf{E} - (\mathbf{k} \cdot \mathbf{E})\mathbf{k} = \frac{\omega^2}{c^2} \mathbf{E} + i\omega \mu_0 \mathbf{J} \qquad (AIV.3)$$

with

$$\mathbf{J} = ne(\mathbf{v}_i - \mathbf{v}_e),$$

give the dispersion relation. However, the equations are rather cumbersome and therefore we will examine special cases.

1. *Longitudinal waves propagating along the magnetic field:* $\mathbf{k} = k\hat{\mathbf{k}}$, $\mathbf{E} = E\hat{\mathbf{k}}$. In this case we have

$$v_{ex} = v_{ey} = 0 \quad \text{and} \quad E_z = +\frac{i\omega m v_{ez}}{e}\left(1 - \frac{c_e^2 k^2}{\omega^2}\right),$$

$$E_z = -\frac{i\omega M v_{iz}}{e}\left(1 - \frac{c_i^2 k^2}{\omega^2}\right).$$

Substituting in (AIV.3) gives

$$0 = \frac{\omega^2}{c^2}E_z - \frac{i\omega\mu_0 n_e^2}{i\omega m}\frac{E_z}{(1 - c_e^2 k^2/\omega^2)} - \frac{i\omega\mu_0 n_e^2}{i\omega M}\frac{E_z}{(1 - c_i^2 k^2/\omega^2)}$$

or

$$1 - \frac{\omega_{pe}^2}{\omega^2 - c_e^2 k^2} - \frac{\omega_{pi}^2}{\omega^2 - c_i^2 k^2} = 0, \qquad \text{(AIV.4)}$$

which for $\omega/k > c_e$ gives

$$\omega^2 \simeq \omega_{pe}^2 + c_e^2 k^2, \qquad \text{(AIV.5)}$$

and for $c_e > \omega/k > c_i$ gives

$$1 + \frac{\omega_{pe}^2}{c_e^2 k^2} \simeq \frac{\omega_{pi}^2}{\omega^2}$$

or

$$\frac{\omega^2}{k^2} = \frac{\omega_{pi}^2 c_e^2}{\omega_{pe}^2 + k^2 c_e^2} = \frac{c_s^2}{1 + k^2 \lambda_D^2}. \qquad \text{(AIV.6)}$$

From (AIV.4) c_e and c_i are seen to be limiting velocities and Figs 6.1 and 7.1 show this graphically.

2. *Transverse waves propagating along field:* $\mathbf{k} = k\hat{\mathbf{k}}$, $\mathbf{E} = e_x\mathbf{i} + E_y\mathbf{j}$. In this case $\mathbf{k} \cdot \mathbf{v} = 0$ and the propagation is unaffected by the inclusion of particle temperatures.

3. *Hybrid waves transverse propagation:* $\mathbf{k} = k\mathbf{i}$, $\mathbf{E} = E_x\mathbf{i} + E_y\mathbf{j}$.

$$-i\omega m v_{ex} = -e(E_x + v_{ey}B_0) + \frac{ik^2 mc_e^2 v_{ex}}{\omega}$$

Then

$$-i\omega m v_{ey} = -e(E_y - v_{ex}B_0).$$

Thus

$$v_{ex} = \frac{e}{m}\frac{\omega_{ce}E_y + i\omega E_x}{\omega_{ce}^2 - \omega^2 + k^2 c_e^2}$$

$$v_{ey} = -\frac{e}{m}\left\{\omega_{ce}E_x + i\omega E_y\left(1 - \frac{c_e^2 k^2}{\omega^2}\right)\right\}\bigg/(\omega_{ci}^2 - \omega^2 + k^2 c_e^2)$$

and

$$v_{ix} = \frac{e}{M} \frac{\omega_{ci} E_y - i\omega E_x}{\omega_{ci}^2 - \omega^2 + k^2 c_i^2}$$

$$v_{iy} = -\frac{e}{M} \left\{ \omega_{ci} E_x - i\omega E_y \left(1 - \frac{c_i^2 k^2}{\omega^2} \right) \right\} \bigg/ (\omega_{ci}^2 - \omega^2 + k^2 c_i^2),$$

giving equations

$$\varepsilon'_{11} E_x + i\varepsilon'_{12} E_y = 0$$

$$\left(\frac{k^2 c^2}{\omega^2} - \varepsilon'_{22} \right) E_y + i\varepsilon'_{12} E_x = 0$$

or

$$\frac{k^2 c^2}{\omega^2} = \frac{\varepsilon'_{11} \varepsilon'_{22} - \varepsilon'^2_{12}}{\varepsilon'_{11}}, \tag{AIV.7}$$

where

$$\varepsilon'_{11} = 1 + \frac{\omega_{pe}^2}{\omega_{ce}^2 - \omega^2 + k^2 c_e^2} + \frac{\omega_{pi}^2}{\omega_{ci}^2 - \omega^2 + k^2 c_i^2},$$

$$\varepsilon'_{22} = 1 + \frac{\omega_{pe}^2 (\omega^2 - k^2 c_e^2)}{\omega^2 (\omega_{ce}^2 - \omega^2 + k^2 c_e^2)} + \frac{\omega_{pi}^2 (\omega^2 - k^2 c_i^2)}{\omega^2 (\omega_{ci}^2 - \omega^2 + k^2 c_i^2)},$$

$$\varepsilon'_{12} = \frac{\omega_{pe}^2 \omega_{ce}}{\omega (\omega_{ce}^2 - \omega^2 + k^2 c_e^2)} - \frac{\omega_{pi}^2 \omega_{ci}}{\omega (\omega_{ci}^2 - \omega^2 + k^2 c_i^2)}.$$

This dispersion relation allows the connection to be made between the Alfvén and ion acoustic branches shown in Fig. 9.3.

4. *Oblique propagation*: $\mathbf{k} = k \cos \theta \hat{\mathbf{k}} + k \sin \theta \mathbf{i}$, $\mathbf{E} = E_x \mathbf{i} + E_y \mathbf{j}$. The form of the terms is such that, where previously $\omega_{c\alpha}^2$ appeared, there now is $\omega_{c\alpha}^2 + k^2 c_\alpha^2$ and this has the effect of causing the 'resonance' at $\omega = \omega_{c\alpha}$ to become a 'resonance' with $\omega/k = c_\alpha$. This observation explains the qualitative differences between the results of Fig. 9.3 and those of Stringer (1963).

It is natural to ask whether the dispersion curves given under different approximations can be conveniently combined into one universal dispersion curve for particular directions of propagation relative to the magnetic field, but, say, combining the electromagnetic, hybrid, and cyclotron modes. This has been considered for longitudinal waves propagating perpendicular to the static magnetic field by Puri, Leuterer, and Tutter (1973) and more generally in a subsequent paper (1975), but is an area which has not as yet been explored thoroughly experimentally in the laboratory.

APPENDIX V

Physical constants and typical values of parameters

e	electron charge	$1 \cdot 602 \times 10^{-19}$ coulomb (C)
m	electron mass	$9 \cdot 109 \times 10^{-31}$ kg
M	ion mass of atomic number A	$1 \cdot 673 A \times 10^{-27}$ kg
k_B	Boltzmann's constant	$1 \cdot 380 \times 10^{-23}$ J K^{-1}
c	velocity of light	$2 \cdot 998 \times 10^{8}$ m s^{-1}
ε_0	permittivity of free space	$8 \cdot 854 \times 10^{-12}$ F m^{-1}
μ_0	permeability of free space	$4\pi \times 10^{-7}$ H m^{-1}
πa_0^2	classical cross-section of hydrogen atom	$0 \cdot 8797 \times 10^{-20}$ m^2
f_{pe}	electron plasma frequency	$\dfrac{1}{2\pi}\left(\dfrac{ne^2}{m\varepsilon_0}\right)^{\frac{1}{2}} = 8 \cdot 978\sqrt{n}$ Hz
f_{pi}	ion plasma frequency	$\dfrac{1}{2\pi}\left(\dfrac{ne^2}{M\varepsilon_0}\right)^{\frac{1}{2}} = 0 \cdot 2095\sqrt{\dfrac{n}{A}}$ Hz
f_{ce}	electron cyclotron frequency	$\dfrac{eB}{2\pi m} = 27 \cdot 99 B$ GHz
f_{ci}	ion cyclotron frequency	$\dfrac{eB}{2\pi M} = \dfrac{15 \cdot 24 B}{A}$ MHz
λ_D	Debye length	$\left(\dfrac{\varepsilon_0 k_B T_e}{ne^2}\right)^{\frac{1}{2}} = 7 \cdot 434\sqrt{\left(\dfrac{T_e}{n}\right)}$(eV) km
c_e	electron thermal speed	$\left(\dfrac{2k_B T_e}{m}\right)^{\frac{1}{2}} = 0 \cdot 5931 \times 10^6 \sqrt{T}$ (eV) m s^{-1}
c_i	ion thermal speed	$\left(\dfrac{2k_B T_i}{M}\right)^{\frac{1}{2}} = 1 \cdot 384 \times 10^4 \sqrt{\left(\dfrac{T}{A}\right)}$ m s^{-1}
c_s	ion sound speed	$\left(\dfrac{k_B T_e}{M}\right)^{\frac{1}{2}} = 0 \cdot 9786 \times 10^4 \sqrt{\left(\dfrac{T}{A}\right)}$ m s^{-1}
c_A	Alfvén speed	$\dfrac{B}{\sqrt{(\rho\mu_0)}} = 2 \cdot 181 \times 10^{16} \dfrac{B}{\sqrt{(nA)}}$ m s^{-1}
	collisionless skin depth	$\dfrac{c}{\omega_{pe}} = 5 \cdot 315 \times 10^6 \dfrac{1}{\sqrt{n}}$ m
ρ_e	electron gyroradius	$\dfrac{c_e}{\omega_{ce}} = 3 \cdot 372 \times 10^{-5} \dfrac{\sqrt{T}}{B}$ m
ρ_i	ion gyroradius	$\dfrac{c_i}{\omega_{ci}} = 1 \cdot 445 \times 10^{-4} \dfrac{1}{B}\sqrt{(TA)}$ m

neutral density at $0°$ C (273 K) and 1 Torr $= 3 \cdot 3 \times 10^{22}$ m^{-3}

REFERENCES

General

EMELEUS, K. G. and WOOLSEY, G. A. (eds) (1970). *Discharges in electronegative gases.* Taylor and Francis, London.

ENGEL, A. von (1965). *Ionized gases.* Oxford University Press.

FRANCIS, G. (1956). *Handb. Phys.* **22**, 53.

HUXLEY, L. G. H. and CROMPTON, R. W. (1973). *The diffussion and drift of electrons in gases.* Wiley, New York.

MASSEY, H. S. W., BURHOP, E. H. S., and GILBODY, H. B. vol. I, 1969; vol. II, 1969; vol. III, 1971; vol. IV, 1974. *Electronic and ionic impact phenomena.* Oxford University Press.

McDANIEL, E. W. and MASON, E. A. (1973). *The mobility and diffusion of ions in gases.* Wiley, New York.

MOTLEY, R. W. (1975). *Q-machines*, Academic Press, New York.

OLESON, N. L. and COOPER, A. W. (1968). *Adv. Electronics Electron Phys.* **24**,

SHKAROFSKY, J. P., JOHNSTON, T. N., and BACHYNSKI, M. P. (1966). *The particle kinetics of plasmas.* Addison-Wesley, Massachusetts.

STIX, T. H. (1962). *The theory of plasma waves.* McGraw-Hill, New York.

SWIFT, J. D. and SCHWAR, M. J. R. (1970). *Electrical probes for plasma diagnostics.* Iliffe, London.

Chapter 1

ALLIS, W. P. (1956). *Handb. Phys.* **21**, 383.

BOYD, R. L. F. and TWIDDY, N. D. (1959). *Proc. R. Soc. A* **250**, 53.

CRAVATH, A. M. (1930). *Phys. Rev.* **36**, 238.

DRESSELHAUS, G., KIP, A. F., and KITTEL, C. (1955). *Phys. Rev.* **98**, 368.

DRUYVESTEYN, M. J. (1930). *Physica, Eindhoven* **10**, 61.

INGRAHAM, J. C. (1965). M.I.T. Res. Lab. Electronics Quart. Prog. Rep. No. 77, p. 112.

McDONALD, A. D. and BROWN, S. C. (1949). *Phys. Rev.* **76**, 1634.

RAYMENT, S. W. and TWIDDY, N. D. (1968). *Proc. R. Soc. A* **304**, 87.

RAYMENT, S. W. and TWIDDY, N. D. (1969). *J. Phys. D* **2**, 1747.

TOWNSEND, J. S. (1925). *J. Franklin Inst.* **200**, 563.

Chapter 2

BOHM, D. (1949). In *The characteristics of electrical discharges in magnetic fields.* (eds A. Guthrie and R. K. Wakerling), McGraw-Hill, New York.

BRYANT, G. H. and FRANKLIN, R. N. (1963). *Proc. phys. Soc.* **81**, 531.

FORREST, J. R. and FRANKLIN, R. N. (1966). *Br. J. appl. Phys.* **17**, 1569.

—— (1969). *J. Phys. B* **2**, 471.

GENTLE, K. W. (1966). *Physics Fluids.* **9**, 2203.

GRANOVSKY, V. (1940). *Dokl. Akad. Nauk. S.S.S.R.* **28**, 37.

HARRISON, E. R. and THOMPSON, W. B. (1959). *Proc. phys. Soc.* **74**, 145.

KILLIAN, T. (1930). *Phys. Rev.* **35**, 1238.

KLARFELD, B. N. (1941). *Fiz. Zh.* **5**, 155.

MEWE, R. (1967). *Proc. 8th Int. Conf. Ioniz. Phenom. Gases, Vienna*, p. 103.
MIERDEL, G. and SCHMALENBERG, W. (1936). *Wiss. Veröff. Siemens-Werken* **15,** 60.
SCHOTTKY, W. (1924). *Phys. Z.* **25,** 635.
SELF, S. A. (1965). *J. appl. Phys.* **36,** 456.
—— EWALD, H. N. (1966). *Physics Fluids.* **9,** 2486.
SWAIN, D. W. and BROWN, S. C. (1971). *Physics Fluids.* **14,** 1383.
TONKS, L. and LANGMUIR, I. (1929). *Phys. Rev.* **34,** 876.
WOODS, L. C. (1965). *J. Fluid Mech.* **23,** 315.

Chapter 3

ALFVÉN, H. and FALTHAMMAR, C-G. (1963). *Cosmical electrodynamics*, p. 204. Clarendon Press, Oxford.
ALLIS, W. P. and GORDON, E. T. (1957). M.I.T. Quart Prog. Rep. No. 46, P. 13; and No. 47, p. 4.
ANDERSON, J. M. (1964). *Physics Fluids.* **7,** 1517.
BARRETT, P. J. (1966). Ph.D. Thesis, University of London.
BICKERTON, R. J. and ENGEL, A. von (1956). *Proc. phys. Soc. B* **69,** 468.
BOHR, N. (1911). Dissertation, Copenhagen.
CHAPMAN, S. and COWLING, T. G. (1939). *Mathematical theory of non-uniform gases*. p. 327. Cambridge University Press. 2nd edn.
CRAWFORD, F. W., EWALD, H. N., and SELF, S. A. (1967). *J. appl. phys.* **38,** 2753.
ECKER, G. and KANNE, H., (1964). *Physics Fluids.* **7,** 1834.
FORREST, J. R. and FRANKLIN, R. N. (1966*a*). *Br. J. appl. Phys.* **17,** 1061.
—— (1966*b*). *Br. J. appl. Phys.* **17,** 1569.
—— (1967). *Proc. 8th Int. Conf. Ioniz. Phenom. Gases, Vienna*, p. 110, 115 (see also B. A. Anicin, p. 175).
—— (1968). *Proc. R. Soc. A* **305,** 251.
FRANCIS, G. (1956). *Handb. Phys.* **22,** 169.
LITTLE, P. F. and JONES, H. G. (1965). *Proc. phys. Soc.* **85,** 979.
ROKHLIN, G. N. (1939). *Fiz. Zh.* **1,** 347.
SELF, S. A. (1967) *Physics Fluids.* **10,** 1569.
STEENBECK, M. (1935). *Wiss. Veröff Siemens-Werken* **15(2),** 1.
VAN LEEWEN, J. H. (1919). Dissertation, Leyden.
VENTRICE, C. A. and BROWN, C. E. (1972). *Br. J. appl. Phys.* **43,** 368.

Chapter 4

ANDREWS, J. G. (1969). D.Phil. Thesis, Oxford University.
—— VAREY, R. H. (1970). *Nature, Lond.* **225,** 270.
—— ALLEN, J. E. (1971). *Proc. R. Soc. A* **320,** 459.
BLANK, J. L. (1968). *Physics Fluids.* **11,** 1686.
BOHM, D., (1949). In *The characteristics of electrical discharges in magnetic fields* (eds A. Guthrie and R. K. Wakerling), McGraw-Hill, New York.
CHILD, C. D. (1911). *Phys. Rev.* **32,** 492.
CRAWFORD, F. W. and CANNARA, A. B. (1965). *J. appl. Phys.* **36,** 3135.
—— —— SELF, S. A. 1967. Stanford University Report No. SU-IPR 185.
FORREST, J. R. and FRANKLIN, R. N. (1968). *J. Phys. D* **1,** 1357.
FRANKLIN, R. N. and OCKENDEN, F. R. (1970). *J. Plasma Phys.* **4,** 371.

FRIEDMAN, H. W. (1967). *Physics Fluids.* **10,** 2053.
—— LEVI, E. (1967). *Physics Fluids.* **10,** 1499.
GOLDAN, P. (1970). *Physics Fluids.* **13,** 1055.
HARRISON, E. R. and THOMPSON, W. B. (1959). *Proc. phys. Soc.* **74,** 145.
LAM, S. H. (1967). *Proc. 8th Int. Conf. Ioniz. Phenom. Gases, Vienna,* p. 545.
LANGMUIR, I. (1913). *Phys. Rev.* **2,** 450.
—— (1920). *Gen. elect. Rev.* **23,** 503 and 589.
MULLER, K. G. and WAHLE, P. (1970). *Z. Naturf.* **25,** 525.
PAK, T. S. and EMELEUS, K. G. (1971). *Proc. 10th Int Conf. Ioniz. Phenom. Gases, Oxford,* p. 361.
PERSSON, K. B. (1962). *Physics Fluids.* **5,** 1625.
PREWETT, P. D. and ALLEN, J. E. (1976) *Proc. R. Soc.* A **348,** 435.
ROSA, R. (1973). *Proc. 11th Int. Conf. Ioniz. Phenom. Gases, Prague,* p. 66.
SHOCKLEY, W. (1949). *Bell Syst. tech. J.* **28,** 435.
SU, C. H. and LAM, S. H. (1963). *Physics Fluids.* **6,** 1479.
TAYLOR, R. J., BAKER, D. R., and IKEZI, H. (1970). *Phys. Lett.* **24,** 206.

Chapter 5

ALLIS, W. P. and ROSE, D. J. (1964). *Phys. Rev.* **93,** 84.
BENSON, F. A. (1965). *Voltage stabilization.* MacDonald, London.
BIONDI, M. A. and BROWN, S. C. (1949). *Phys. Rev.* **76,** 1697.
BRYANT, G. H. (1966). Ph.D. Thesis, London University.
CARUSO, A. and CAVALIERE, A. (1964). *Br. J. appl. Phys.* **15,** 1021.
EBERT, W. (1975). *Proc. 12th Int. Conf. Ioniz. Phenom. Gases, Eindhoven,* p. 219.
EDGLEY, P. D. (1975). D.Phil. Thesis, Oxford University.
ELENBAAS, W. (1951). *The high pressure mercury vapor discharge,* North Holland, Amsterdam.
FRANKLIN, R. N. (1963). *Proc. 6th Int. Conf. Ioniz. Phenom. Gases, Paris,* p. 157.
GOLUBOWSKI JU. B., KAGAN, JU. M., and MICHEL, P. (1971). *Beitr. Plasmaphys.* **11,** 121.
GRAY, E. P. and KERR, D. E. (1962). *Ann. Phys.* **17,** 276.
HAYESS, E. and WOJACZEK, K. (1970). *Beitr. Plasmaphys.* **10,** 407.
HEYMANN, P. (1969). *Beitr. Plasmaphys.* **9,** 491.
HOLM, R. (1932). *Z. Phys.* **75,** 171.
ILIC, D. B. (1973). *J. appl. Phys.* **44,** 3993.
IRISH, R. T. and BRYANT, C. H. (1964). *Proc. phys. Soc.* **84,** 975.
KAGAN, YU. M. and LYAGUSCHENKO, R. T. (1962). *Sov. Phys. tech. Phys.* **7,** 535
—— (1964). *Optics Spectrosc., Wash.* **17,** 90.
KENTY, C. J. (1938). *J. appl. Phys.* **9,** 765.
LYNCH, R. H. (1967). *J. appl. Phys.* **38,** 3965.
PREWETT, P. D. (1975). D.Phil. Thesis, Oxford.
PRINS, M. and SMITS, P. M. M. (1975). *Proc. 12th Int. Conf. Ioniz. Phenom. Gases, Eindhoven,* p. 67.
REIMANN, H. and HEYMANN, P. (1970). *Beitr. Plasmaphys.* **10,** 417.
RILEY, J. R. (1970). *J. Phys.* D **3,** 1226.
SABADIL, H. (1973). *Beitr. Plasmaphys.* **13,** 236.
SEELIGER, R. (1949). *Ann. Phys.* **6,** 93.
SMITH, D., DEAN, A. G., and ADAMS, N. G. (1974). *J. phys.* D **7,** 1944.
SMITS, R. M. M. and PRINS, M. (1975). *Proc. 12th Int. Conf. Ioniz. Phenom. Gases, Eindhoven,* p. 68.

STANGEBY, P. C. and ALLEN, J. E. (1971). *J. Phys. A* **4**, 108.
—— (1973). *J. Phys. D* **6**, 224.
STEENBECK, M. (1939). *Wiss-Veröff Siemens-Werken.* **18**, 318.
THOMPSON, J. B. (1959). *Proc. R. Soc. A* **262**, 503.
VALENTINI, H. B. (1972). *Beitr. Plasmaphys.* **12**, 87.
—— (1974). *Beitr. Plasmaphys.* **14**, 201.
VENZKE, D., HAYESS, E., and WOJACZEK K. (1966). *Beitr. Plasmaphys.* **6**, 365.
WEAVER, L. A. and FREIBERG, R. J. (1968). *J. appl. Phys.* **39**, 4283.
WOJACZEK, K. (1966), *Beitr. Plasmaphys.* **6**, 211.
—— (1967). *Beitr. Plasmaphys.* **7**, 149.
—— (1968). *Beitr. Plasmaphys.* **8**, 109.
—— (1969). *Beitr. Plasmaphys.* **9**, 243.
WOOLSEY, G. A., EMELEUS, K. G., GRAY, E. W., and COULTER, J. R. M. (1967). *Int. J. Electron.* **22**, 235.

Chapter 6

AKAO, Y. and IDA, Y. (1963). *J. appl. Phys.* **34**, 2119.
—— (1964). *J. appl. Phys.* **35**, 2565.
BALDWIN, D. E. (1969). *Physics Fluids.* **12**, 279.
—— IGNAT, D. (1969). *Physics Fluids.* **12**, 697.
BARBERIO-CORSETTI, P. (1970). Princeton Plasma Physics Laboratory Report MATT-773.
BERNSTEIN, I. B. (1958). *Phys. Rev.* **109**, 10.
BHATNAGAR, P. L., GROSS, E. P., and KROOK, M. (1954). *Phys. Rev.* **94**, 511.
BOHM, D. and GROSS, E. P. (1949). *Phys. Rev.* **75**, 1851.
BRYANT, G. H. and FRANKLIN, R. N. (1963). *Proc. phys. Soc.* **81**, 531.
BUCHSBAUM, S. J. and HASEGAWA, A. (1964). *Phys. Rev. Lett.* **12**, 615.
COQUIL, E. L., HENRY, D., LEMEUR, J. P., CASTRAC, C., and TREGUIER, J. P. (1971). *Revue Phys. Appl.* **6**, 467.
CRAWFORD, F. W. (1963). *Phys. Lett.* **5**, 244.
—— (1967). *Proc. 8th. Int. Conf. Ioniz. Phenom. Gases, Vienna.* Invited Papers, p. 109.
CRAWFORD, F. W. and TATARONIS, J. A. (1965a). *J. appl. Phys.* **36**, 2930.
—— (1965b). *Int. J. Electron.* **19**, 557.
CRAWFORD, F. W. and HARKER, K. J. (1972). *J. Plasma Phys.* **8**, 261.
DATTNER, A. (1961). *Proc. 5th Int. Conf. Ioniz. Phenom. Gases, Munich,* vol. II, p. 1477.
DIAMENT, P. (1967). *Physics Fluids* **10**, 470.
ECKER, G. and FROMLING, G. (1974). *Z. Naturf.* **29a**, 1863.
FERGUSON, E. A. (1971). ORNL Report No. TM-3610.
FRANKLIN, R. N. (1964). *J. Electron Control.* **17**, 513.
—— (1968). *Plasma Physics* **10**, 805.
—— HAMBERGER, S. M., LAMPIS, G., and SMITH, G. J. (1975). *Proc. R. Soc. A* **347**, 1.
FRIED, B. D. and CONTE, S. D. (1961). *The plasma dispersion function,* Academic Press, New York.
FRISCH, H. L. and PEARSON, G. A. (1966). *Physics Fluids.* **9**, 2467.
GAUTSCHI, W. (1961). *SIAM J. numer. Anal.* **7**, 187.
GOULD, R. W. (1960). *Bull. Am. phys. Soc.* **5**, 322.
GRANATSTEIN, V. L., KORN, P., OJO, A., and SCHLESINGER, S. P. (1967). *J. appl. Phys.* **38**, 1969.

HENRY, D. and TREGUIER, J. P. (1972). *J. Plasma Phys.* **8,** 311.

HOW, J. A. and BLEVIN, H. A. (1976). *J. Phys. D.* **9,** 1123.

HUGGINS, R. W. and RAETHER, M. (1966). *Phys. Rev. Lett.* **17,** 745.

IGNAT, D. W. (1970). *Physics Fluids.* **13,** 1771.

JACKSON, E. A. and RAETHER, M. (1966). *Physics Fluids.* **9,** 1257.

LANDAU, L. (1946). *Fiz. Zh.* **10,** 25.

LEUTERER, F. (1969). *Plasma Phys.* **11,** 618.

MALMBERG, J. H. and WHARTON, C. B. (1964). *Phys. Rev. Lett.* **13,** 184.

NICKEL, J. C., PARKER, J. V., and GOULD, R. W. (1964). *Physics Fluids.* **1,** 1489.

O'NEIL, T. M. (1965). *Physics Fluids.* **8,** 2255.

PEARSON, G. A. (1966). *Physics Fluids.* **9,** 2454.

PERATT, A. L. (1973). *Physics Fluids.* **16,** 1032.

PRINZLER, H. (1969). *Plasma Phys.* **11,** 507.

ROOS, B. W. (1969). *Analytic functions and distributions in physics and engineering.* Wiley, New York.

SCHMITT, H. J., MELTZ, G., and FREYHEIT, P. J. (1965). *Phys. Rev.* **139A,** 1432.

SPITZER, L. (1956). *Physics of fully ionized gases.* Interscience, New York.

STIX, T. H. (1962). *The theory of plasma waves.* McGraw-Hill, New York.

TANAKA, S., KUBO, H., and MITANI, K. (1965). *J. phys. Soc. Japan.* **20,** 462.

TRIVELPIECE, A. W. and GOULD, R. W. (1959). *J. appl. Phys.* **30,** 1784.

TUTTER, M. (1968). *Plasma Phys.* **10,** 775.

VANDENPLAS, P. E. (1968). *Electron waves and resonances in bounded plasmas.* Wiley, London.

Chapter 7

ALEXEFF, I., JONES, W. D., and MONTGOMERY, D. (1968). *Physics Fluids.* **11,** 167.

ARUNASALAM, V. and BROWN, S. C. (1965). *Phys. Rev.* **140A,** 471.

ATKINSON, H. H. (1963). *Proc. 4th Int. Conf. on Microwave Tubes, Schiveningen, The Netherlands,* p. 55. Centrex, Eindhoven.

BARRETT, P. J. (1966). Ph.D. Thesis, London University.

BEKEFI, G. (1966). *Radiation processes in plasmas.* Wiley, New York.

BUZZI, J. M. (1974). *Physics Fluids.* **17,** 716.

CAVALIERE, A., ENGELMANN, F., and SESTERO, A. (1970). *Physics Fluids.* **11,** 158.

DAVIES, D. E. (1966). D.Phil. Thesis, Oxford University.

EMELEUS, K. G. (1964). *Adv. Electronics Electron Phys.* **20,** 78 et seq.

ESTABROOK, K. and ALEXEFF, I. (1972). *Physics Fluids.* **15,** 2026.

EWALD, H. N., SELF, S. A., and CRAWFORD, F. W. (1969). *Physics Fluids.* **12,** 303.

FENNEMAN, D. B., RAETHER, M. and YAMADA, M. (1973). *Physics Fluids.* **16,** 871.

GOLDAN, P. D. and LEAVENS, W. M. (1970). *Physics. Fluids.* **13,** 433.

ILIC, D. B., WHEELER, G. M., CRAWFORD, F. W., and SELF, S. A. (1974). *J. Plasma Phys.* **12,** 433.

LITTLE, P. F. (1961). *Proc. 5th Int. Conf. Ioniz. Phenom. Gases, Munich,* vol. II, p. 1440.

—— HAMBERGER, S. M. (1966). *Nature, Lond.* **209,** 972.

PREWETT, P. D. and ALLEN, J. E. (1976). *Proc. R. Soc. A* **348,** 435.

SATO, N., SASAKI, A., AEKI, K., and HATTA, Y. (1967). *Phys. Rev. Lett.* **19,** 1174.

TANACA, H., HIROSE, A., and KOGANEI, M. (1967). *Phys. Rev.* **161,** 94.

THONEMANN, P. C. (1955). *Appl. scient. Res. B* **5,** 237.

TWOMEY, D. and FRANKLIN, R. N. (1974). *J. Phys. D* **7,** 1963.

WADA, J. Y., KNECHTLI, R. C. and HEIL, H. (1966). *Proc. 7th Int. Conf. Ioniz. Phenom. Gases, Belgrade,* vol. I, p. 424.

WEYNANTS, R. R., MESSAIEN, A. M., and VANDENPLAS. P. E. (1973). *Physics Fluids.* **16,** 1962.

WOODS, L. C. (1965). *J. Fluid Mech.* **23,** 315.

Chapter 8

BARRETT, P. J. and LITTLE, P. F. (1965). *Phys. Rev. Lett.* **14,** 356.

DUNCAN, A. J. and FORREST, J. R. (1971). *Physics Fluids.* **14,** 1973.

FRANKLIN, R N. (1970). In *Physics of ionized gases.* (ed. B. Navinsek), Herceg Novi.

FRIED, B. D., WHITE, R. B., and SAMEC, T. K. (1971). *Physics Fluids,* **14,** 2388.

GARSCADDEN, A. and LEE, D. A. (1966). *J. electron. Control.* **20,** 567.

—— BLETZINGER, P., and SIMENON, T. C. (1969). *Physics Fluids.* **12,** 1833. **12,** 1833.

GENTLE, K. W. (1966). *Physics Fluids.* **9,** 2203, 2212.

GRABEC, I. and MIKAC, S. (1974). *Plasma Phys.* **16,** 1155.

GRANOWSKI, W. L. (1955). *Der Elektrische Strom in Gas.* Akademie-Verlag, Berlin.

HAAS, R. A. (1973). *Phys. Rev.* **A8,** 1017.

NAKAMURA, M., ITO, M., NAKAMURA, Y., and ITOH, T. (1975). *Physics Fluids.* **18,** 651.

NIGHAN, W. L. and WIEGAND, W. J. (1974). *Phys. Rev.* **A10,** 922.

PEKAREK, L. (1971). *Proc. 10th Int. Conf. Ioniz. Phenom. Gases.* Invited Papers, p. 365.

—— KREJCI, V. (1963). *Czech. J. Phys.* **13,** 881.

—— MASEK, K., and ROHLENA, K. (1970). *Czech. J. Phys.* **B20,** 879.

PUPP, W. (1935). *Phys. Z.* **33,** 273.

SATO, M. (1973). *Beitr. Plasmaphys.* **13,** 9.

SWAIN, D. W. and BROWN, S. C. (1971). *Physics Fluids.* **14,** 1383.

TWOMEY, D. and FRANKLIN, R. N. (1974). *J. Phys. D.* **7,** 1963.

WOJACZEK, K. (1971). *Beitr. Plasmaphys.* **11,** 335.

Chapter 9

ALLEN, T. K., BAKER, W. R., PYLE, R. V., and WILCOX, J. M. (1959). *Phys. Rev. Lett.* **2,** 283.

AULT, E. R. and IKEZI, H. (1970). *Physics Fluids.* **13,** 2874.

BAYNHAM, A. C. and BOARDMAN, A. D. (1971). *Plasma effects in semiconductors.* Taylor and Francis, London.

CHRISTOPOULOS, C., BOSWELL, R. W., and CHRISTIANSEN, P. J. (1974). *Phys. Lett.* **47A,** 239.

DAVIES, B. (1969). *Phys. Rev. Lett.* **22,** 1246.

—— CHRISTIANSEN, P. J. (1969). *Plasma Phys.* **11,** 987.

DELLIS, A. N. and WEAVER, J. M. (1963). *Proc. phys. Soc.* **83,** 473.

DRUMMOND, J. E. (1968). *Physics Fluids.* **11,** 1196.

EDGLEY, P. D., FRANKLIN, R. N., HAMBERGER, S. M., and MOTLEY, R. W. (1975). *Phys. Rev. Lett.* **34,** 1269.
FERRARI, R. L. and KLOZENBERG, J. P. (1968). *J. Plasma Phys.* **2,** 283.
GALLET, R. H., RICHARDSON, J. M., WIEDER, B., WARD, G. D., and HARDING, G. N. (1960). *Phys. Rev. Lett.* **4,** 347.
HIROSE, A., ALEXEFF, I., and JONES, W. D. (1970). *Physics Fluids.* **13,** 2039.
JEPHCOTT, D. F. and STOCKER, P. M. (1962). *J. Fluid. Mech.* **13,** 587.
KLOZENBERG, J. P., MCNAMARA, B. and THONEMANN, P. C. (1965). *J. Fluid. Mech.* **21,** 545.
LEE, J. C., FESSENDEN, T. J., and CRAWFORD, F. W. (1969). *Proc. 9th Int. Conf. Ioniz. Phenom. Gases, Bucharest,* p. 475.
LEHANE, J. A. and THONEMANN, P. C. (1965). *Proc. phys. Soc.* **85,** 301.
—— PAOLONI, F. J. (1972). *Plasma Phys.* **14,** 701.
MALEIN, A. (1965). *Nucl. Fus.* **5,** 352.
MCKAY, D. H. (1967). D.Phil. Thesis, Oxford University.
MORROW, R. and BRENNAN, M. H. (1971). *Plasma Physics.* **13,** 75.
PAOLONI, F. J. (1973). *Plasma Phys.* **15,** 475.
SATO, N., SUGAI, H., SASAKI, A., and HATAKEYAMA, R. (1974). *Physics Fluids.* **17,** 456.
SCHMITT, J. P. (1972). *Physics Fluids.* **15,** 2057.
—— (1973*a*). *Plasma Phys.* **15,** 677.
—— (1973*b*), *Phys. Rev. Lett.* **31,** 982.
SWANSON, D. G., GOULD, R. W., and HERTEL, R. H. (1964). *Physics Fluids.* **7,** 269.
WOODS, L. C. (1962). *J. Fluid Mech.* **13,** 570.

Chapter 10

ALCOCK, M. W. and KEEN, B. E. (1971). *Phys. Rev.* **A3,** 1087.
BOHM, D. (1949). In *The characteristics of electrical discharges in magnetic fields.* (eds A. Guthrie and R. K. Wakerling), McGraw-Hill, New York.
CHEN, F. F. (1967). *Physics Fluids.* **10,** 1647.
CHU, T. K., COPPI, B., HENDEL, H. W., and PERKINS, F. W. (1969). *Physics Fluids,* **12,** 203.
HOH, F. C. and LEHNERT, B. (1960). *Physics Fluids.* **3,** 600.
KADOMTSEV, B. B. (1965). *Plasma turbulence.* Academic Press, New York.
—— NEDOSPASOV, A. V. (1960). *J. nucl. Energy C* **1,** 230.
KEEN, B. E. and ALDRIDGE, R. V. (1970). *Plasma Phys.* **12,** 839.
LITTLE, P. F. and MIDDLETON, C. R. (1969). *Nucl. Fus.* **9,** 67.
MIKHAILOVSKI, A. N. (1965). *Rev. Plasma Phys.* Consultants Bureau: **3,** 159.
MONTICELLO, D. A. and SIMON, A. (1974). *Physics Fluids.* **17,** 791.
NAGAI, Y., IMAZU, S., MARUYAMA, T., and MARUYAMA, T. (1975). *J. phys. Soc. Japan* **38,** 295.
POWERS, E. J. and MUMOLA, P. B. (1971). *Plasma Phys.* **13,** 817.
RUKHADZE, A. A. and SILIN, V. P. (1969). *Sov. Phys. Usp.* **11,** 659.
SIMON, A. and SHIAU, J. N. (1969). *Physics Fluids.* **12,** 2630.
TIMOFEEV, A. V. T. (1964). *Sov. Phys. tech. Phys.* **8,** 862.

Appendices

FRANKLIN, R. N. and OLDFIELD, M. L. G. (1969). *Int. J. Electron.* **27,** 431.
PURI, S., LEUTERER, G., and TUTTER, M. (1973). *J. Plasma Phys.* **9,** 89.
—— (1975). *J. Plasma Phys.* **14,** 169.

STRINGER, T. E. (1964). *Plasma Phys.* **6,** 267.

VAN DUYN, C. J., PRINS, M., and SMITS, R. M. M. (1975). *Proc. 12th Int. Conf. Ioniz. Phenom. Gases, Eindhoven* p. 69.

VRIENS, L. (1973). *J. appl. Phys.* **44,** 3980.

WOJACZEK, K. (1967). *Proc. 8th Int. Conf. Ioniz. Phenom. Gases, Vienna.* Invited Papers, p. 413.

W58030—Plasma phenomena.

INDEX